Power MOSFET Design

Power MOSFET Design

B. E. Taylor

International Rectifier Co (GB) Ltd
UK

JOHN WILEY & SONS
Chichester · New York · Brisbane · Toronto · Singapore

Copyright © 1993 by by John Wiley & Sons Ltd,
Baffins Lane, Chichester,
West Sussex PO19 1UD, England

All rights reserved.

No part of this book may be reproduced by any means,
or transmitted, or translated into a machine language
without the written permission of the publisher.

Other Wiley Editorial Offices

John Wiley & Sons, Inc., 605 Third Avenue,
New York, NY 10158-0012, USA

Jacaranda Wiley Ltd, G.P.O. Box 859, Brisbane,
Queensland 4001, Australia

John Wiley & Sons (Canada) Ltd, 22 Worcester Road,
Rexdale, Ontario M9W 1L1, Canada

John Wiley & Sons (SEA) Pte Ltd, 37 Jalan Pemimpin #05-04,
Block B, Union Industrial Building, Singapore 2057

Library of Congress Cataloging-in-Publication Data

Taylor, B. E.
    Power Mosfet design / B. E. Taylor.
      p. cm.
    Includes bibliographical references and index.
    ISBN 0 471 93802 5
      1. Metal-oxide semiconductor field-effect transistors—Design.
    2. Power transistors—Design.   I. Title.
    TK7871.95.T39 1993
    621.3815′284—dc20   92–44450
                                                               CIP

British Library Cataloguing in Publication Data

A catalogue record for this book is available from the British Library

ISBN 0 471 93802 5

Typeset in 11/12½ Palatino by International Scientific Communications, Amesbury, Wilts
Printed and bound in Great Britain by Biddles Ltd, Guildford & Kings Lynn

Contents

Introduction ix
Acknowledgements xi

1 MOS-GATED TRANSISTORS (MGTs): THEIR STRUCTURE
   AND IN-CIRCUIT BEHAVIOUR 1
   1.1 BJT control charge (Q_{BJT}) 2
   1.2 MGT gate control charge (Q_{gate}) 3
   1.3 Various MGT structures and their effects 5
   1.4 Conductivity modulation 10
   1.5 IGT latching 16
   1.6 Gate oxide thickness and electrostatic damage 18

2 GATE CONTROL 23
   2.1 Current limiting and voltage clamping by gate control 41

3 OVER-VOLTAGE PROTECTION 43
   3.1 Active clamps 44
   3.2 Avalanche 46
   3.3 Passive clamps 47
   3.4 Soft capacitor clamps 52
   3.5 Capacitive clamps and symmetric push–pull converters 55
   3.6 Transient voltage suppressors 56
   3.7 Voltage dependent resistors 57
   3.8 Transient voltage suppressor diodes 57
   3.9 Gate control 58
   3.10 Circuit layout 60
   3.11 Topological clamping 61

4 OVER-CURRENT PROTECTION 63
   4.1 Current mirror/sensing MOSFETs 64
   4.2 Circuit techniques for over-current protection 69
   4.3 Current measurement 69
   4.4 Turn-on snubbers 73

5 THERMAL MANAGEMENT — 77

- 5.1 $R_{ds[on]}$ variation as a function of temperature — 79
- 5.2 $V_{ce[sat]}/V_{cs[on]}$ variation as a function of temperature — 80
- 5.3 Other thermal management considerations — 80
- 5.4 Using multiple junctions (parallel connection) — 84

6 EMI/RFI AND LAYOUT — 87

- 6.1 Techniques for reducing interference at source — 89

7 GENERAL CIRCUIT TECHNIQUES — 99

- 7.1 Using medium voltage MOSFETs in high voltage circuits — 99
- 7.2 Uninterruptible power supplies — 101
- 7.3 Emitter switching with bipolar junction transistors — 102
- 7.4 Series connection of MGTs and slave control of their gates — 105
- 7.5 Ultimate switch speed, its ramifications and attainment — 106
- 7.6 Using standard MOSFETs in high frequency PWM inverters — 108
- 7.7 High frequency operation of IGTs — 110
- 7.8 MOSFET switching aid to IGT for HF switching — 111
- 7.9 Parallel connection of MGTs — 111

8 POWER SUPPLIES — 115

- 8.1 Linear regulators — 116
- 8.2 Low drop-out high efficiency linear regulators — 117
- 8.3 Switching regulators — 121
- 8.4 Choice of frequency — 121
- 8.5 Topology preferences and their impact on power switch technology — 122
- 8.6 Single switch forward/flyback converter — 122
- 8.7 Symmetric push–pull — 124
- 8.8 Bridge topologies — 125
- 8.9 600 V BV_{DSS} MOSFETs in 220 V off-line flyback supplies — 125
- 8.10 Effective turns ratio variation — 131
- 8.11 MGTs as rectifiers in power supplies — 133

9 MOTOR DRIVES AND CONTROLLERS — 139

10 AUTOMOTIVE ELECTRONICS — 151

- 10.1 Chassis electronics — 152
- 10.2 Body electronics — 163

11 ELECTRONIC BALLASTS — 169

- 11.1 Effects of harmonic content of drive current waveform — 171

12 AUDIO AMPLIFICATION 179
12.1 Parallel operation of MOSFET in the linear region of operation 180
12.2 Switching audio power amplifiers 185
12.3 The audio power ampliverter 187

Appendix SAFE OPERATING AREA FAILURES IN BJTs 191

Index 197

Introduction

Over the years several books have been published relating to subjects as diverse as the design of power converters, motor control, cycloconverters and others, all of which may be regarded as being interrelated. The books covering the subject of power converters are mainly concerned with the design of Switch Mode Power Supplies (SMPS). Of late, virtually no subject material has been provided relating to the linear power supply.

The vast majority of the books also cover the use of Bipolar Junction Transistors (BJTs) and this subject material has been backed-up by a large quantity of papers and application notes. However, the abundance of all of this literature relating to the BJT is, in many instances, irrelevant to today's technological requirements.

The introduction, over a decade ago, of the vertical geometry co-planer double-diffused power MOSFET (D-MOS) resulted in the first power semiconductor becoming available to seriously rival the dominant position of the high-voltage fast switching bipolar transistor. The virtual acceptance of the power MOSFET in today's market as a commodity item would have been expected to provide literature which would cater for a very wide spectrum of users and potential users.

Unfortunately, an insignificant number of papers and books are concerned with the actual design philosophy related to circuits which utilise power MOSFETs. When MOSFETs are addressed the range of cover is usually found to be severely restricted. It is not restricted in the sense that the depth of coverage is limited, but in that the material is either aimed specifically at undergraduate students and 'academics' or is so low key in its wording as to be relevant solely to secondary school students and hobbyists. Almost no literature is available which attempts to span the whole range of readership, and to include the practising engineer and technician. This book is an attempt to correct this omission.

The subjects covered are mainly concerned with the driving and protection of all power MOS-gated structures — from the conventional power MOSFET device to those with current-sensing capability and the ones with in-built conductivity modulation. A degree of familiarity with chronologically earlier power transistors (BJTs) will be beneficial to the reader, but not essential. A thorough understanding of the basics related to the subject of power electronics is necessary if a complete grasp of the material is to be achieved. This understanding should also encompass, if possible, the subject of magnetics as perceived today.

INTRODUCTION

I make no apology for the inclusion of some of my favourite circuits. I am well aware that I am leaving myself open to the accusation of being egotistic. My firm belief is that the interest factor of these circuits will be found to stimulate the imagination of the reader. If this stimulation, in turn, fosters the desire towards further experimentation on the part of the reader, then inclusion of these circuits will have proven to have been worthwhile.

The inclusion of some circuits may well be questioned, since these circuits could well be demonstrated as being capable of functioning with BJTs or any other type of semiconductor power switch. It will become obvious that I have included these circuits because of their superior behaviour when they are found to be designed around the power MOS-gated transistor.

Acknowledgements

I am deeply indebted to all my friends and colleagues who have shared their knowledge, wisdom and free time in advising me on the best way to approach the writing of this book.

I would also like to thank all those who have attended the many seminars that I have had the privilege to present. It is to some of those attendees that I am particularly grateful since it was their continued suggestion for me to publish the tips which I gave at the seminars which has resulted in me putting pen to paper.

I am also deeply appreciative of the understanding and encouragement I have received from my family and most especially my wife, who was certainly not aware of the consequences to herself, from the encouragement she gave. It was her unstinting editing and re-typing that has made this publication possible.

1
MOS-gated Transistors (MGTs): Their Structure and In-circuit Behaviour

It should be made clear to the reader that there is no intention to discuss the *static induction transistor* and the *static induction thyristor* in this book. The reason for omitting these devices is that they do not belong to the family of MOS-gated transistors and more importantly they have been found to be virtual laboratory specimens. The only non-MGT device which I shall, on occasion, refer to is the Bipolar Junction Transistor (BJT), because of certain similarities between the two parts.

There is one area where the MOSFET (and nearly all other MOS-gated transistors for that matter) is vastly inferior to the conventional BJT. The overwhelming superiority of the BJT lies in its circuit symbol.

The power MOSFET's overall structure should be well known by now, since it has been discussed at great length already; and therefore will not be discussed in any significant detail within this book.

It is sufficient to say that once the basic device has been fabricated then continuation of the fabrication of the control structure (the gate) is achieved by covering the exposed upper surface of the device with an insulating material (in this case it is usually silicon dioxide). The gate structure is now deposited upon this insulating medium and after suitable etching to a predetermined (mask) pattern it is covered by another layer of the same insulator material. The fabrication of the Insulated Gate Transistor (IGT) is achieved by a back diffusion into the drain structure of a MOSFET of a minority carrier injection layer. In the case of an N-channel device this injection layer will be a P-type semiconductor; and N-type for a P-channel structure. This last description is an over-simplification of the process for fabricating an IGT. This is especially true of the newer IGTs which have been recently introduced as second generation parts.

It can therefore be seen that, between the gate and the remainder of the MOSFET's structure, there exists a very high resistance. This intrinsically high resistance between gate and the rest of the structure results in the MOSFET having a very high input resistance (which is often erroneously termed as being a high input impedance). This high input resistance gives the MOS-gated transistor a number of important differences to the BJT. One of these differences will be scrutinised, since this difference is significant enough to quantify the present day popularity of the power MOSFET; and will be of equal importance to the IGT.

The power MGT and the BJT are both characterised as being fundamentally charge controlled. The similarity both starts and ends at this point, in that the BJT's charge is inherently current derived whereas that of the MOSFET is fundamentally voltage derived.

It is now necessary to examine the profound nature of the two devices' charge control characteristics.

1.1 BJT CONTROL CHARGE (Q_{BJT})

Over a relatively short proportion of the total conduction time, of the BJT, the charge can be expressed as:

$$Q_{BJT} = I_b t_{on} \tag{1.1}$$

where I_b is the average value for the base current flowing for the period of time t_{on} (which may be defined as the total switch-on time of the BJT and is the sum of turn-on delay time and current rise time). This charge can be further enlarged to be:

$$Q_{BJT} = I_c / h_{FE} t_{on} \tag{1.2}$$

where I_c is the collector current (averaged over t_{on}), and h_{FE} is the static current transfer ratio of the BJT. If the duration of t_{on} is of the order of a few hundred nanoseconds or even as long as one microsecond, then the use of the static current transfer ratio is perfectly valid. It is also interesting to note that the base charge is proportional to collector current.

The energy (E_{Base}) expended in accumulating this charge can be expressed as:

$$E_{Base} = V_{be}(I_c / h_{FE}) t_{on} \tag{1.3}$$

where V_{be} is the average value of base emitter voltage of the transistor for t_{on}. V_{be} is found to be proportional to the product of the base spreading resistance and the emitter current. The implication therefore is that V_{be} is proportional to the collector current. It can be demonstrated that V_{be} has a logarithmic relationship with the collector current. This logarithmic tendency is of little relevance since for all practical purposes a power transistor will never be used at extremely small values of collector current where the logarithmic effect is most noticeable.

As an indication of the overall drive efficiency it is not sufficient merely to discuss the base control charge. It is equally necessary to discuss the power dissipated in the base and this can be expressed as follows:

$$P_{base} = (E_{base} \times \text{frequency}) + P_{Base}(t_{cond})$$

where t_{cond} is conduction time of the BJT and is that part of the total on-time which does not include the turn-on time.

If

$$t_{ON} = t_{on} + t_{cond}$$

then

$$P_{base} = V_{be}(I_c/h_{FE})t_{ON}f \qquad (1.4)$$

Returning to the subject of transistor control charge it becomes essential to discuss the control charge of the MGT.

1.2 MGT GATE CONTROL CHARGE (Q_{gate})

The MGT gate control charge can be expressed as:

$$Q_{gate} = C_{iss}V_{gs} \qquad (1.5)$$

where C_{iss} is the input capacitance of the MGT and V_{gs} is the gate to source voltage of the device. The reader is reminded that C_{iss} is the sum of two capacitances — namely the sum of C_{GS} (the gate to source capacitance) which is a real capacitance; and C_{DG} (the drain to gate capacitance) which is a voltage dependent capacitance. This value of gate to source voltage is a function of the type of device (whether it is a standard or logic level MOSFET or IGT) and the voltage required to achieve full enhancement. It should be noted that the drain current is not mentioned.

Note that the gate charge is difficult to calculate owing to the voltage dependency of C_{DG} and is best measured.

Of greater significance is the energy (E_{Gate}) expended in accumulating this charge which can best be expressed as:

$$E_{Gate} = 0.5 C_{iss} V_{gs}^2 \qquad (1.6)$$

Once again, for the purpose of indicating the overall drive efficiency, it is necessary to discuss the power required to drive the gate and this can be expressed as:

$$P_{Gate} = E_{Gate} \times \text{frequency} \qquad (1.7)$$

After expansion equation (1.7) becomes:

$$P_{gate} = 0.5 C_{iss} V_{gs}^2 f \tag{1.8}$$

In the case of P_{gate} this power would be dissipated within the gate structure only if the gate structure resistance happened to be significantly higher than the drive circuit impedance. P_{base}, on the other hand, is always dissipated within the transistor structure of the BJT itself. It should be remembered that the magnitude of control power is not important in itself. The real importance of this power is the indication of the drive circuit's power delivery and handling capability.

Having derived equations (1.4) and (1.8), it would be beneficial if the effects can be demonstrated practically. Consider two well known examples of both types of semiconductor: the ubiquitous 2N3055 (BJT) and IRF140 (MOSFET). Both devices will be used in a simple circuit to switch a resistive load at 50% duty cycle and at a frequency of 1 kHz. For simplicity the load current for both devices can be set at 10 A.

For the 2N3055:

$$V_{be} = 1.5 \text{ V}, \quad h_{FE} (@ \ I_c = 10 \text{ A}) = 5$$

Using equation (1.4) we have

$$P_{base} = [\{0.75 \times (1 \times 1E^{-6})\} + \{1.5 \times (10/5) \times 499E^{-6}\}] \times 1000$$
$$= \mathbf{1.49\,775 \text{ W}}$$

For the IRF140:

$$V_{gs} = 10 \text{ V}, \quad C_{iss} = 1600 \text{ pF}$$

Using equation (1.8) we have

$$P_{Gate} = \tfrac{1}{2} \times 1600E^{-12} \times 10^2 \times 1000 = \mathbf{80 \ \mu W}$$

The disparity between the bipolar and the MOSFET is self-evident. It is fair to acknowledge that it can be demonstrated that the BJT can be improved by converting its structure into a monolithic Darlington; but it is equally fair to say that the improvement would not equate to the drive efficiency demonstrated for the MOSFET.

Just as the preceding paragraphs have been used to highlight the difference between a particular BJT and a particular MOSFET, so it can be demonstrated that equal discrepancies also exist between MOSFETs from one supplier to another, and is not merely confined to the differences between *lateral* and *vertical* structures.

1.3 VARIOUS MGT STRUCTURES AND THEIR EFFECTS

A parasitic BJT is found within every MOSFET structure. This parasitic is easily discernible in Figure 1.1. In the case of the IGT this parasitic component becomes a thyristor. This parasitic thyristor does have the capability of latching on, especially if the IGT is incorrectly fabricated or used.

The possibility of the IGT latching is totally undesirable and the consequences to circuits may well be catastrophic. Techniques to prevent the latching of IGTs will be discussed later. From both of the adjacent cell structures of Figure 1.1, the two-junction transistor (NPN) structures are clearly in evidence. The base–emitter junction of the parasitic transistor within the lateral MOSFET structure indicates that the base of this NPN is open-circuit (it is connected externally to the N emitter), whereas the base–emitter junction of the vertical structure is found to be partially short-circuited by the source metallisation. Extensive detail of the oxide and metallisation is not shown. Very much more detail is given in Figure 1.4.

Examination of the equivalent circuit in Figure 1.2 will show that there is a resistor between the base and the emitter of the parasitic NPN.

The value of this base resistor (R_{be}), is dependent upon several factors — such as the base spreading resistance, the diffusion profile, the concentration of the dopant and the thickness of the source/emitter metallisation. The value for R_{be}, and how low it is will determine how good is the definition of the overall structure.

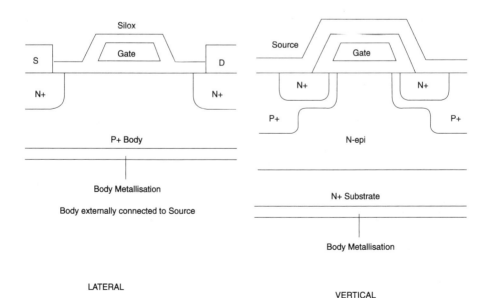

Figure 1.1 Illustration showing structure of MOSFET cell

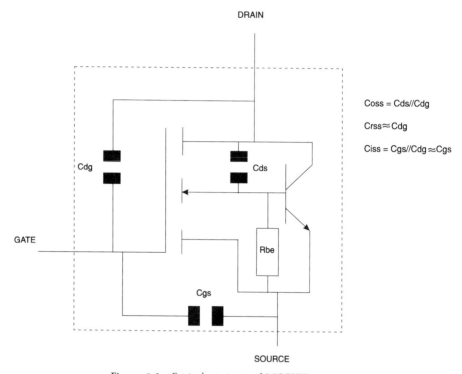

Figure 1.2 Equivalent circuit of MOSFET structure

If R_{be} was shown to be exactly equal to zero, then the only form of break-down which could be sustained by the NPN would be V_{cbo}. Any finite value of resistance (R_{be}) above zero would imply that the break-down characteristic will be modified and that at some value of avalanche current the base–emitter junction would become forward biased and that emitter injection would commence, resulting in the parasitic BJT turning on and entering into second breakdown.

The control of R_{be} and its final value usually indicates some measure of the ruggedness of the MOSFET. The ruggedness of MOSFETs will be found to vary from one manufacturer to another.

When considering the type of MOSFET to be included within any design, the designer should scrutinise the requirements of the design in its entirety. The performance of the chosen device will ultimately be reflected in the total system cost, which in turn will not only include the purchase price of components.

It is of little consequence if the build price of a particular piece of equipment happens to be the lowest in the market-place, if the same equipment proves to be completely unreliable. The cost of warranty service, and possible component replacement, could conceivably negate any profit which may have been accrued. Above all else will be the effect upon the goodwill and esteem of the end-user of the equipment.

1.3 VARIOUS MGT STRUCTURES AND THEIR EFFECTS

The reliability of any piece of equipment can only reflect the equality which has been built into it.

The first golden rule, that will now be stated, should read: *Never skimp on the quality of components.*

It should be obvious that application of this rule should be tempered with some caution.

The designer may well imagine that he/she is forced to walk a tightrope over the decision of quality. This is only partly true, since there are also useful 'trade-offs' which may be made: trade-offs that will enable the use of components reflecting the highest standards of quality.

An excellent example of how 'Trade-offs' can be made is to consider how the use of rugged MOSFETs can lead to real savings in the overall cost of equipment.

Consider the use of a half-bridge topology in an off-line Switched Mode Power Supply (SMPS). In certain instances it would be mandatory to install 500 V (BV_{DSS}) MOSFETs as the power switches. But, virtually identical overall performance can be achieved by the judicious use of 400 V MOSFETs which are endowed with avalanche capability. The 400 V device should have the same $R_{ds[on]}$ as its 500 V counterpart but utilise a device that is possibly 25–50% smaller than its higher voltage sibling.

Proof of this assumption can readily be provided within the following paragraph.

The rectified d.c. bus voltage for a worst case off-line supply is approximately equal to 375 V (UK specification of 240 ± 10%. It is usual to include a margin of safety, in order that certain International Safety Legislation requirements be complied with. The safety margin will almost certainly include transient over-voltage capability due to stray inductances. In certain instances a double helping of safety is introduced by the incorporation of surge suppressors across the d.c. bus.

It is now prudent to examine all of the assumptions to determine where, if any, further savings may be made for a supply of 500 W of throughput power.

(1) It will be assumed that a RFI/EMI filter is found to be necessary. It is customary for such a filter to include an inductor in series with both line and neutral connections. It is also customary to have a capacitor connected from line to neutral immediately after the line and neutral inductors.

(2) If output hold-up is a specification requirement of the power supply then the capacitance of the reservoir capacitor post rectification will be considerable. The capacitance value for this capacitor can be calculated from:

$$C_{Res} = 2P_{out}/\varepsilon(t_{hold\text{-}up}/(V_1^2 - V_2^2) \tag{1.9}$$

8　MOS-GATED TRANSISTORS (MGTs)

where $t_{\text{hold-up}}$ is the hold-up time for the output (and is usually quoted for 'half a missing mains cycle' or may for a high specification supply even cater for a 'a whole missing mains cycle' — a part of the specification that is sometimes called the Brown-out specification), P_{out} is the output power of supply, V_1 is the minimum d.c. bus voltage (derived from the supply's input specification), and V_2 is the d.c. bus voltage where output regulation can no longer be maintained, C_{Res} is the capacitance of the reservoir capacitor and ε is the target efficiency of the supply.

For the 500 W supply as (defined earlier) with a hold-up time of 38 ms (one missing 50 Hz cycle); and with an efficiency of 75% and using equation (1.9) it can be shown for $C_{\text{Reservoir}}$ to have a capacitance of 1320 μF.

The energy which this capacitor can absorb without exceeding the BV_{DSS} of the MOSFET can be deduced from:

$$E_{\text{clamp}} = C_{\text{Res}} (V_{\text{DS}}^2 - V_{\text{hl}}^2)/2 \tag{1.10}$$

where E_{clamp} is the energy required to raise the voltage across the reservoir capacitor C_{Res} from the UK high-line voltage of 375 V d.c. to $V_{\text{DS}} = 400$ V, the clamped safe value of BV_{DSS}.

For a high-line voltage of 264 V a.c., and the reservoir capacitance calculated by using equation (1.9) above the clamp energy can be demonstrated to be greater than 12 joules. This energy capability is in excess of the published safety legislation specifications and shows that 400 V MOSFETs can be utilised provided one further precautionary measure is also undertaken.

This measure is to ensure that the total voltage stress which is applied across the MOSFET does not exceed BV_{dss}. This voltage stress will usually be generated by the load and also by any stray inductance within the circuit.

One possible method that is frequently employed is the use of *current snubbers*. The use of current snubbers is advocated across the load (which in this instance will usually be a transformer). The use of other snubbers may, on the other hand, prove to be unwarranted.

The maximum transient voltage owing to stray inductances can be determined from:

$$V_{\text{transient}} = BV_{\text{dss}} - V_{\text{hl}}$$

where V_{hl} is the high-line voltage and has been demonstrated to be equal to 375 V for the d.c. bus. $V_{\text{transient}}$ is regarded as being equal to the integral of L_{stray} and the rate of change of current through the stray inductance L_{stray}.

The value of L_{stray} should be minimised as a matter of principle but can be determined (empirically) as being equal to 10 nH in value for every inch of excess chassis wiring or length of printed circuit-board track.

1.3 VARIOUS MGT STRUCTURES AND THEIR EFFECTS

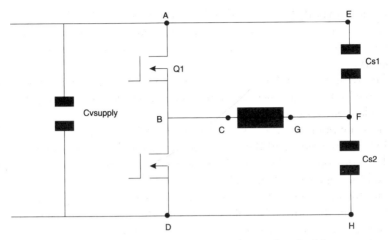

Figure 1.3 Schematic of 500 W SMPS showing length of L_{stray}

Since both the rate of change of current (at switch-off) and the total value of stray inductance (effected at switch-off) are contributory to the level of $V_{transient}$; and if the use of snubbers — around the power switch(es) — is deemed to be undesirable, then one of two possible alternatives should be considered: (a) the rate of change of current (at switch-off) should be reduced or (b) the energy content of the transient should be reduced.

Reducing the switch-off speed of the switch (and thereby reducing di/dt) will be discussed in Chapter 2; therefore only the latter alternative will be explored at this point.

In the circuit of the proposed 500 W power supply the total current path-length should be ascertained with the empirical value of 10 nH per inch being used to calculate the value of L_{stray}. Assume that Q1, in the circuit of Figure 1.3, is one of the switches. The total current path-length is the sum of the lengths from: (1) $C_{VSUPPLY}$ to point A, (2) point A to the drain of Q1, (3) the source of Q1 to point B, (4) point B to point C, (5) point A to point E, (6) point E to C_{S1}, (7) C_{S1} to point F, (8) point F to point G, (9) point F to C_{S2}, (10) C_{S2} to point H, (11) point H to point D, and finally point D back to $C_{VSUPPLY}$. The load which is shown connected between points C and G is the transformer. $C_{VSUPPLY}$ is used to decouple the d.c. bus, and should have as low an inductance content as possible. By inference the current-path should be minimal. Having deduced the energy content of the transient the reader should now examine the repetitive avalanche energy (E_{AR}) capability of the MOSFET; if E_{AR} (and I_{AR} — the avalanche current capability) both exceed the in-circuit values, and provided the power dissipation P_d is not exceeded, then the use of further snubbers (around the power switch) can be regarded as being superfluous.

If all of the points outlined above are rigorously observed, it should be obvious that the rugged, high quality, MOSFET enables the user to utilise a device of smaller size than that accommodated within the less well endowed part, while at the same time the elimination of the cost (and the cost of mounting) of the additional snubbers can be regarded as being cost effective without sacrificing reliability.

The examination of MOSFET structures so far has highlighted their ease of drive, whilst the ruggedness of some (but not all) devices has displayed their cost reduction capabilities.

The assumption so far has been that the MOSFET is also more efficient than the BJT. For the most part this is true for medium voltage applications requiring products of 500 V or less in rating, or where the frequency of operation is beyond the capability of the BJT.

Unfortunately, during the period of its introduction to the power electronics market-place, the power MOSFET was subjected to every possible means of scrutiny by the public at large. Some of the scrutiny from opponents to the technology took the form of criticism. In certain instances the criticism was indeed valid, whereas some of the methods employed in the presentation of these objections were questionable in the extreme.

The major objection of any consequence was the undeniably high value of conduction losses sustained by the true high-voltage power MOSFET, when operated in 'saturation'. The reason for the unacceptable on-state losses was the high value of $R_{ds(on)}$. The 'unacceptable' value of $R_{ds(on)}$ was mostly applicable to break-down voltages of 600 V or higher.

The BJT, to which the power MOSFET was frequently compared, was endowed with the ability to be operated in a condition termed *deep saturation*. This condition is also referred to sometimes as *hard saturation*. Operating in hard saturation resulted in the BJT sustaining unquestionably lower on-state losses than could be achieved by the medium and high voltage MOSFETs.

The ability (of the BJT) to operate in the condition known as hard saturation is achieved by the further condition known as conductivity modulation.

1.4 CONDUCTIVITY MODULATION

When the base of a BJT is over-driven, namely the device transits from operating in its quasi-saturated condition to operating in the hard saturation region (the current gain changes from the nominal h_{FE} value to that defined by the conditions specifying $V_{ce(sat)}$). An excess charge of minority carriers is injected into the base region. The net result of this action (of over-driving) is to increase the flow of holes, in the case of an NPN transistor, from base to collector; with a corresponding increase in electron current-flow from collector to emitter, whilst not incurring the penalty of increased on-state voltage. This condition which is also known as deep saturation will manifest itself by the indication of V_{be} becoming greater than V_{ce}. This forward-biasing of the collector–base junction can under certain conditions of drastic over-drive be observed at the external

1.4 CONDUCTIVITY MODULATION

connections of the transistor. Normally forward-biasing of the collector–base junction may be measured at terminals on the junction itself.

The price which must be paid for this superior 'on-state efficiency' is the very considerable impairment of the BJT to switch-off at anywhere near its potential speed capability.

The MOSFET, being a pure *majority carrier* device, suffers from a complete lack of ability in being able to modulate its conductivity.

The outcome of the initial conflict in the market-place has been a virtual abandonment, by the BJT, of some of its traditional sockets — such as the SMPS — to the 'up-start' technology. The BJT's new market, according to some vendors, is intended as being the high-voltage traction application. This statement does not imply that BJTs are not to be found in SMPS sockets. The implication is that ever-increasing switch frequencies will inevitably mean the near total abandonment of these sockets by the BJT.

One of the developments which took place with the conventional MOSFET structure was a back diffusion of acceptor atoms into the drain epi layer, effectively resulting in the creation of a P-type layer in series with the drain of the MOSFET.

Such a modified structure is depicted in Figure 1.4 alongside the layout of the conventional MOSFET structure as a means of comparing the two types of structure. The reader is cautioned to be aware that the structures as depicted in Figure 1.4 are merely a means of representing the types of structure. They do not indicate diffusion profiles or depths.

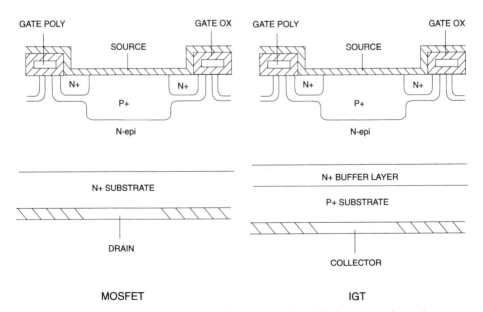

Figure 1.4 Sectional view of a conventional MOSFET and modified structure of a conductivity modulated device

The cross-sectional views of the two types of power MOS structure are illustrated in Figure 1.4. The shaded areas marked source and emitter are the respective metallisation. This also applies to the drain and collector metallisation.

The inclusion of the P-type layer of the new structure now permits the injection of minority carriers (holes in the case of an N-channel MOSFET drain) into the N+ buffer layer, further enhancing the flow of electrons, and thereby effectively reducing the 'on-state resistance' $R_{DS[on]}$ of the device. The inclusion of the N+ buffer layer is to facilitate the collection of minority carriers at turn-off, thereby enhancing the turn-off performance. N+ buffer layers may be used, on their own, as an alternative to normal carrier lifetime control or may be used as a feature which is in addition to lifetime control.

The various generic names attributed to the new MOS-gated device structure are indeed diverse. These names include:

(1) COMFET from **CO**nductivity **M**odulated **FET**.
(2) GEMFET from **G**ain **E**nhanced **MOSFET**.
(3) IGT from **I**nsulated **G**ate **T**ransistor.
(4) IGBT from **I**nsulated **G**ate **B**ipolar **T**ransistor.

There is no fundamental difference between the basic structure — although some vendors may infer the existence. It is certain that there will be variations in overall performance, but this may be attributed to fine tuning.

For the purpose of further discussion, throughout the remainder of this book, I propose the usage of the name IGT. This choice does not arise out of any particular preference on my part — it just so happens to be the shortest of all of the abbreviations.

Although the difference between the two structures may appear to be marginal the reader should not be lulled into any false sense that the effect upon performance has not been profound.

The original MOSFET structure included a two-junction parasitic element (a BJT), where this element constituted the body-drain diodes; this misnomer, in turn is a carry-over from the lateral MOSFET structure. The body-drain diode is effectively connected in anti-parallel to the MOSFET, and is usually schematically depicted in the schematic symbol.

This anti-parallel diode within the MOSFET structure can be considered as being a mixed blessing. In certain applications where a commutating diode in an anti-parallel connection is required — and where the reverse recovery speed/time t_{rr} is of little concern — the body-drain diode will be regarded most definitely as an asset, since it comes free. Where the t_{rr} is important the parasitic diode is now found to be a genuine liability unless due allowance is made for its characteristics. Some of the causes of the variability of t_{rr} in the parasitic diode are t_j and BV_{dss} of the MOSFET.

1.4 CONDUCTIVITY MODULATION

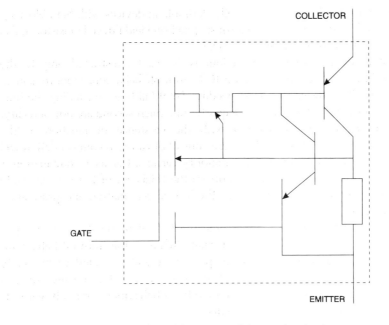

Figure 1.5 Equivalent circuit of the new bipolar device

A later chapter will cover fabrication and circuit techniques which are used to overcome any of the deficiencies of the body-drain diode and so render it usable.

Careful scrutiny of the cross-sectional view and at the equivalent circuit of the IGT reveals the lack of the gift of the body-drain diode of the MOSFET. Instead, one is made immediately aware of the presence of a three-junction parasitic (a thyristor).

The equivalent circuit of the new IGT structure is given in Figure 1.5 and the reader is urged to pay due regard to the existence of the three junctions, since this is one of the reasons for the device on occasion being named a bipolar. There is no real inference of this device being another variant of BJT. It is used as a means of being able to differentiate between a device whose flow of current involves no junction at all (the MOSFET), and with the other case where minority carrier injection is desired. The extra P-layer of the IGT, although enhancing the device with the ability to modulate its conductivity and hence lower its 'on-state losses', has the dubious distinction of altering the switching behaviour, of the original MOSFET structure, to a significant extent.

The reader should pay special attention to the manufacturer's data sheet limits, since any deviation outside of the prescribed limits, relating to applied reverse voltage, could result in failure to the parasitic and hence to the IGT. Failure to observe data sheet limits could also result in this parasitic element latching on, with potentially disastrous results.

Conductivity modulation is especially desirable in devices with high blocking voltage to reduce the effects of the intrinsic Junction Field Effect Transistor (JFET) in the drain of the conventional MOSFET.

The alteration in switching behaviour is, in reality, restricted only to the turn-off behaviour. Second generation IGTs turn-off behaviour is more akin to that of a MOSFET turning off less speedily. Total fall times (including the time for the tail-current reaching zero) of a quarter of a microsecond are now a reality.

This can be directly connected with the existence of junction 3 (the emitter–base junction of the PNP) where control of carrier recommendation will be determined by the lifetime of the minority carriers. Carrier recombination is controlled entirely by a combination of: (a) the thickness of the P+ substrate and therefore its resistivity, and (b) to the lack of termination and passivation of the junction.

It is theoretically feasible to overcome the lack of termination and to gain access to this junction and thereby externally influence the 'turn-off behaviour' of this junction and the IGT itself. In practice it will be found to be nearly impossible (and this is mainly a question of economics) for the user to gain access to this junction and to therefore modify its behaviour externally with the use of active and/or passive components.

It is therefore only the vendor who can control the behaviour of the device. In doing this the vendor usually fabricates the device to achieve an overall compromise. The turn-off behaviour is achieved by trading-off conduction efficiency by the use of lifetime control of the minority carriers.

Examination of the idealised waveforms in Figure 1.6 will indicate that the turn-on behaviour of the IGT is fundamentally the same as that of the familiar and conventional MOSFET. The turn-on delay time t_{dON} is almost identical to that of a MOSFET. The V_{ce} waveform exhibits none of the severe *collector voltage tailing* (into hard saturation) which is a fundamental trait of the BJT. The reader should note that t_{Vtail} is moderately short. It is acknowledged that there is some 'voltage tailing' compared with the conventional MOSFET, whose voltage waveform is the finely hatched line which is given for the purpose of comparison.

Please note the use of BJT terminology for the voltage and current rise and fall times. These parameters have been somewhat neglected with the MOSFET since the values have been within the control of the user.

In the case of the IGT they do become significant, especially towards the prediction of the turn-on and turn-off energies (losses).

The turn-off behaviour of first generation IGTs is in some respects totally alien to that of the normal MOSFET.

When the gate voltage transits from the cessation of the V_{ce} *rise time* phase the collector current enters into and progresses through the initial *fall-time* phase until the gate voltage reaches the threshold value as shown by the horizontal hatched line, which passes through the V_{gs} waveform.

1.4 CONDUCTIVITY MODULATION

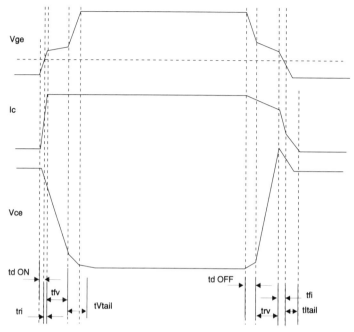

Figure 1.6 Ideal IGT waveforms

The initial collector current fall-time is as rapid as that of the FET. It is at the V_{gs} threshold transition that the similarity ends. The collector current now enters a region of definite current tailing — which can also be observed in the switching behaviour of some badly driven BJTs.

The collector current tailing phenomenon is not akin to that of a badly driven BJT but is entirely due to *carrier recombination* which takes place in an undefined manner within the base of the PNP structure. In the case of the BJT, manifestation of current-tailing could be an imminent indication of the onset of second breakdown.

Scrutiny of the turn-off portion of the waveforms of Figure 1.6 will indicate that the period of current tailing is a region of high dissipation during the switching cycle.

The magnitude and duration of the tail is dependent upon several variables; all of them being beyond the control of the user. The duration of the tail (and to a much lesser extent the magnitude) can be altered by any one of several methods employing *carrier life-time control*. The more usual of these techniques involve either doping with a heavy metal, ion implantation or electron irradiation. Any one of these techniques will enhance the turn-off behaviour but usually at the expense of increased saturation losses.

All of the foregoing information relating to conductivity modulation is not intended to deter the reader from using these devices. They are merely given in order to forewarn the user of the device's overall capability. It will in time become apparent that the IGT's overall capability is in fact formidable.

The higher *current density* of the IGT, compared with the medium voltage MOSFET and the BJT, results in better silicon utilisation for medium to high voltage applications. The draw-back is that IGT use is limited to operating frequencies of 50 kHz for quasi square-wave switching and several hundred kilohertz for quasi-resonant operation. There are drive techniques available that will increase the square-wave frequency limit considerably, but the economics of these techniques are in the main relatively questionable. It could conceivably be more efficient and cost effective merely to parallel connect standard MOSFETs, in order to reduce the conduction losses. In a later chapter a useful circuit will be demonstrated that will enable 250 kHz square-wave operation to be economically viable.

1.5 IGT LATCHING

In normal use the IGT will usually be operated within its *continuous collector current* rating (I_c). It could well be that for certain low duty cycle applications the designer may well wish to work with a significantly higher value of collector current than the continuous rating. In this case it is perfectly acceptable to use the *pulsed collector current* rating (I_{CM}) provided that the average value of collector current remains within the continuous rating value.

I have frequently been asked if the value of I_{CM} can be exceeded for extremely short intervals in time, provided that the $t_{j(max)}$ value is not exceeded. The value of I_{CM} and also for I_{DM} with the standard MOSFET is quoted as being pulse-width limited, within the bounds of t_j. The argument, frequently offered by unwary users, is that extremely short duration pulses which are within the *thermal time constant of the silicon mass* should be permissible — provided the magnitude and duration of the pulse is kept within the limit of the I^2t fusing capability of the wire bonds.

The preceding paragraph is given for the benefit of those readers who may well accept the vendor's data sheet publicity related to *latch-free* operation. It should be remembered that this applies to operation within data sheet limits of I_c and $t_{j(max)}$.

Early IGTs suffered from a propensity to latch when operating at elevated temperatures, even though these elevated junction temperatures were rigorously maintained within the manufacturer's limits. The reason for this tendency to latch was fundamentally due to poor process control of *diffusion profiles* and also of emitter metallisation. The result was excessive leakage current into the P-base of the parasitic NPN leading to *emitter injection* and ultimately to regenerative turn-on of the parasitic thyristor.

1.5 IGT LATCHING

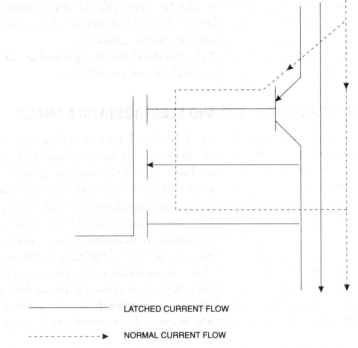

———————— LATCHED CURRENT FLOW

- - - - - - - - - - ▸ NORMAL CURRENT FLOW

Figure 1.7 Current flow in the IGT shown for both normal and latched operation

The caution that I am advocating, as regards excessive collector current, arises out of the manifestation of an entirely different phenomenon which is fundamentally not linked to leakage and to operating junction temperature directly.

The description of one method of inducing emitter injection, as given below, is indirectly affected by junction temperature and its consequential variation of on resistance.

Figure 1.7 illustrates the path of normal current flow, which is shown as the two broken lines in the equivalent circuit. Note that a major percentage of the current within the device flows through the PNP structure.

Flow of this current through the inversion layer (MOSFET channel), shown shaded, will create a potential gradient across the channel.

Since the channel is in direct contact with the P-base of the NPN, the peak of the ohmic drop in the channel will result in a certain value for V_{be} (for the NPN) developing.

If this V_{be} is of sufficient magnitude to induce emitter injection in the NPN, the NPN will then turn on and divert current away from the MOSFET inversion layer. This diversion of the current away from the channel is seen to be regenerative and results in the parasitic thyristor latching. The latched current path is shown, for the purposes of simplicity, as the solid line flowing through the PNP.

All of the above explanation is directly applicable to the conventional MOSFET, also; but in this instance the parasitic NPN's turn on leads to secondary break-down within the parasitic, owing to current crowding.

The golden rule associated with IGTs can therefore be expressed as: *Observe data sheet limits, especially those of the pulsed collector current* (I_{CM}).

1.6 GATE OXIDE THICKNESS AND ELECTROSTATIC DAMAGE

A device having a maximum V_{gs} of ± 20 V will have a typical gate oxide thickness in the region of 1000 Å units. In the case of the logic-level type device this oxide layer will be found to have been severely curtailed to about half of the original thickness. It should be noted that these two oxide layer thicknesses do not apply to devices from all of the various manufacturers. Logic-level parts, with an oxide layer thickness of 1000 Å, are perfectly viable today. The reason why vendors opt to use the reduced gate oxide thickness is that common die sizes, cell geometries and common mask sets can be used during wafer fabrication.

Slight differences in doping may have to be made in conjunction with the reduction in oxide thickness. The advantages to this approach are twofold. First, $R_{DS[ON]}$ and I_D characteristics also tend to be common between standard gate parts and their logic-level counterparts. Second, development costs are greatly reduced.

When considering the use of any type of MOS (Metal Oxide Semiconductor) device, the user should always be aware of the gate structure having limited toleration to any degree of abuse. Those who fail to observe this warning will find that their circuits will suffer from significant unreliability.

The 500 Å thickness increases the necessity for maintaining vigilance over possible gate over-voltage, while at the same time introducing further special considerations. These extra precautions arise out of the effect the oxide thickness can have upon reliability and how this effect becomes manifest during initial fabrication.

At the time of wafer fabrication a certain amount of surface contamination of the wafer is inevitable. This contamination occurs even with the maintenance of the most stringent precautions of cleanliness within the diffusion facility. The contaminant ions collect along the surface of the gate polysilicon and not as a cloud as is shown in Figure 1.8. I have depicted these ions as a cloud in order to highlight the position they migrate to.

The surface contamination is in the form of *ions* which may be deposited upon the non-metallised surface, i.e. only the first layer of gate oxide has been deposited upon the surface of the water. The polysilicon gate structure will then be fabricated, and the final layer of oxide insulator deposited over the surface of the water. After etching the surface metallisation will complete the fabrication of the silicon. The contaminant ions are now firmly trapped within the overall structure.

1.6 GATE OXIDE THICKNESS AND ELECTROSTATIC DAMAGE 19

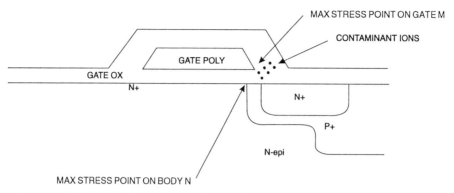

Figure 1.8 Gate structure of MGT showing maximum stress points

Application of a voltage to the gate terminal will cause the ions to migrate to the point of maximum stress, namely the points of maximum stress are given as point M and N in Figure 1.8. The resultant *ion migration* exacerbates the stress condition by creating an electric field of extremely high intensity, in the vicinity of the trapped ions. If the intensity of the field, at the point of stress, is of sufficient magnitude, punch-through of the gate oxide becomes almost certain.

The *punch-through* or *rupture* voltage between gate and source is significantly higher than the maximum figure quoted in manufacturer's data sheets, thus providing a useful safety margin. The reader is strongly advised to adhere to the recommended limits for the sake of reliability.

The increase in the percentage of enhancement which may be achieved, due to an increase in gate-source voltage (up to the recommended limit), is negligible and the reader should avoid the application of such voltages.

If the user is locked into the loop of scratching for every small reduction in $R_{DS[ON]}$, then the logical alternative is to use a die of larger area or to parallel connect two or more devices. It can be demonstrated that either of the two alternative solutions will invariably prove to be more cost-effective in the long run.

Another reason for avoiding $V_{gs(max)}$ is the real improvement in expected reliability. Although not readily apparent, it is also an inevitable consequence of ion migration and the resulting failures can be plotted as a function of life expectancy for varying levels of gate voltage and temperature.

Accumulated data covering this aspect of reliability provides overwhelming evidence that violation of this parameter is inexcusable. The user is cautioned against accepting publicity purporting to novel processes that offer vast improvements. The process that is used by one manufacturer is virtually identical to all of the others. The only secret ingredient that is sacrosanct is tight quality control over the whole process.

Data — displaying expected time to failure as a function of operating junction temperature, and plotted for a range of values for V_{gs} — provides sample

MOS-GATED TRANSISTORS (MGTs)

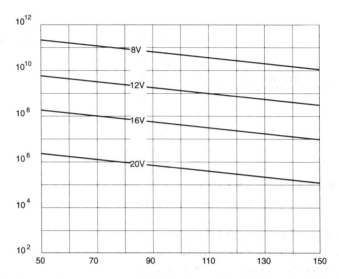

Figure 1.9 Typical time to accumulated 1% failure (Courtesy of International Rectifier Corporation)

evidence of the very real improvement in expected device longevity, as long as V_{gs} is kept at a moderate level.

Simple extrapolation of the curves, followed by a short programme of calculation, will provide convincing evidence that a V_{gs} of 10 V and a maximum t_{case} of 100 °C will provide the gate structure with a life expectancy of 900 000 years approximately, before failures of 1% may be expected. Merely raising V_{gs} to 20 V, while keeping all other conditions constant, will reduce the expected life of the gate structure to a mere 90 years. This reduction in expected life, by approximately 4 orders of magnitude, is illustrated in Figure 1.9.

The golden rule which emerges from the discussions relating to the integrity of the gate structure is: *Exceed the maximum value of V_{gs}, at your peril.*

The maximum operating voltages which I would recommend for the various gate structures approximate to 10 V for a 1000 Å technology device, and 5 V for a 500 Å thickness part. It is permissible to exceed these values only under circumstances that relate to small operating duty cycles.

A good example would be the drive circuit for a single diode injection laser. In this circuit the duty ratio of on to off time is very small, while the peak to average current ratio of drain current is high. In this instance it would be perfectly acceptable to use a relatively small device and to pulse the gate up to the recommended maximum limit in order to prevent the MOSFET from going into its current limit mode of operation.

1.6 GATE OXIDE THICKNESS AND ELECTROSTATIC DAMAGE

Although the reader of this book would normally not become involved in equipment production, handling precautions are strongly recommended. The handling precautions necessary for laboratory conditions need not be as stringent as those applied to a production environment. Normal common-sense precautionary measures should prove to be perfectly adequate. Advice should be sought if there is any doubt about prevailing conditions in an extremely dry environment.

2
Gate Control

In spite of the intention, on my part, to avoid contentious and controversial statements related to comparisons between BJTs and MGTs, I nevertheless find that it has been unavoidable. There is one important point that should be made clear concerning the similarity of the products and refers to the two devices being charge controlled.

In the case of the BJT the charge controlling constituent happens to be current which must be injected into, or withdrawn from, the base of the BJT. For the MGT this charge control variable happens to be a voltage which must be applied to the gate. The implication of this difference in similarity is that the MGT is immeasurably easier to drive and to control than the BJT ever was and could be. This restatement of the view in Chapter 1 can be simplified to read: *it is very much easier to slow down a MOSFET than it is to speed-up the BJT*. First generation IGTs prove to be a very slight exception to this rule, and this last statement will be clarified later in this chapter. The reference to the speed of the MOSFET is not merely wishful thinking. Driven correctly the MOSFET is capable of creating di/dt's of the order of 10 000 A per μs quite easily.

It is only in exceptional circumstances that the MOSFET will ever be required to switch at the true speed of which it is fully capable. In order to fully realise the speed potential of the MOSFET the user must be prepared to circumvent the pitfalls which could be created by the very speed which is being pursued.

The first-time user of the MOSFET, or the person accustomed to using BJTs, may inadvertently try to extract as much speed (which can be sustained by the circuit) as is possible with a view to reducing the total switching losses. This has been one of the main reasons for the choice of the MOSFET as the switching component. Unfortunately this approach could lead to oscillatory transients upon voltage and current waveforms at best, or to disastrous consequences at worst.

These same perturbations, if generated within circuits embodying poor layout, may well create untenable levels of EMI/RFI, which will then have to be filtered out — thus increasing the cost of the circuit and/or affect the overall reliability.

The unwanted transients could lead to premature failure in the MGT and also be the cause of failure to an allied component. With the correct degree of gate control the generation of these transients can be prevented with relative ease. This arises out of the predictable nature of the MOSFET's overall switching behaviour. The same ease will be found to be fundamental to the switch-on behaviour of the IGT. The switch-off behaviour of the IGT is found to differ somewhat to the turn-off behaviour of the MOSFET. Second generation IGTs have turn-off characteristics similar to the better BJT and also to a modestly driven MOSFET.

The reader could be forgiven for inquiring how it would be possible to design a power circuit with predictable switching characteristics, since this would require a considerable degree of control over several variables. The answer is surprisingly fundamental in both theoretical and practical circumstances. The same approach will enable a first-cut approach at performance prediction as an overall measure of efficiency if it is so wished. This results from the determination of switching time that can yield knowledge relating to the turn-on and turn-off switching energies per pulse.

The data sheets of most suppliers of MGTs should include data related to values of *gate-charge* relevant to every one of the MGTs within the catalogue. The values of gate-charge should relate to the three major capacitances of the device.

If this data is not supplied by the manufacturer then my advice to the user would be to seek an alternative source of supply. Another approach which could be adopted would be to inform the vendor that the use of the product will only be investigated upon submission of this information.

The reader may well enquire as to the use of capacitance parameters as a possible alternative. The advice I would give is that these parameters will only provide a crude approximation as to the real performance of the device.

It should be remembered that charge may simply be expressed as the product of current and time. This over-simplification is true for batteries and other accumulators when time may of itself be large. It can be equally valid for small periods in time which normally exist as the switching transitions in electronic circuitry. When it is expressed as an equation the value of Q (the charge) can be shown as:

$$Q = it \qquad (2.1)$$

The ramifications of this simplified equation are extremely profound. If the circuit designer knows the gate-charges which are required by the power device and can design the circuit in such a manner as to utilise and to exploit these charge characteristics, by judicious control of the current flowing into or out of the gate terminal, it will then follow that, time being the ratio of charge and current, the switching times can be predicted with a surprisingly high degree of accuracy and certainty.

Once accustomed to the use of this data, the reader will find that examination of any further data, for a new device, will rapidly enable him/her to predict the switching behaviour and hence the switching times for that particular device.

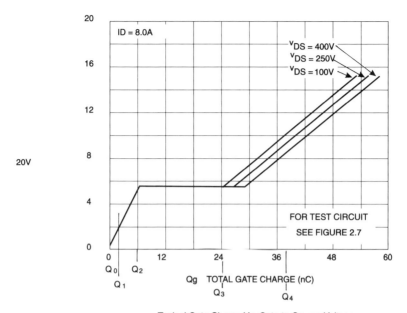

Figure 2.1 Typical gate charge versus V_{gs} curve for IRF840 (Courtesy of International Rectifier Corporation)

Let us consider the curves in Figure 2.1, and see how the theory may be applied.

Upon examination, the curve of gate-charge versus gate to source voltage will illustrate the simplicity in being able to predict the switching behaviour and hence times for a given MOSFET. The horizontal distance from points Q_0 to Q_1 (being a value of charge) can be equated to the time required to raise the voltage of the gate–source capacitance to the threshold voltage for a given flow of current into the gate terminal. This interval is defined as the *turn-on delay time* t_{dON}. Similarly the horizontal distance between points Q_1 and Q_2 can be equated as the *drain current rise-time* t_{ri}. The plateau region of the gate–charge curve is where the *reverse transfer* or *Miller capacitance* becomes active and the drain to source voltage V_{DS} starts to fall. Therefore the horizontal distance between points Q_2 and Q_3 can be defined as being the *drain–source voltage fall-time* (t_{fv}). Finally, the distance from point Q_3 to Q_4 is the excess charge required (and can therefore be equated to time) to fully enhance the channel. This interval can therefore be regarded as being commensurate with the *voltage tail-time*.

Experienced users of power BJTs should be familiar with the phenomenon of *voltage tailing*. This term is applied to the interval between when the transistor originally reaches quasi-saturation and then slowly proceeds to the condition of

26 GATE CONTROL

hard/deep saturation. Voltage tailing is one of the unspoken but none the less contributory causes to the switching losses sustained by a BJT, and usually has resulted in significant discrepancies between calculated dissipation and that determined by thermal measurement. The equivalent effects of voltage tailing in the MGT are significantly less pronounced and may usually be neglected.

It should be noted that the turn-on energy dissipated within the MOSFET will be triangulated and if the variables of supply/clamp voltage and load current are known then the simple formula relating to the area of a triangle will yield the value for this energy.

Virtually the same procedure, but working in reverse, yields *turn-off delay time* (t_{dOFF}) for point Q_4 to point Q_3. Point Q_3 to point Q_2 yields *voltage rise time* (t_{rv}), while point Q_2 to point Q_1 yields *current fall time* (t_{fi}). The charge for going from point Q_1 to point Q_0 is the excess charge required to ensure the MOSFET is definitely off. It does not equate to any particular time definition which is applicable to the BJT.

The reverse procedure for yielding times will also therefore yield the turn-off energy per pulse which is dissipated within the MGT.

All of the foregoing text relating to gate-charge and times is for circuits where the power device's load is meant to be a constant current generator. For most practical purposes the everyday load for this purpose will contain some form of inductive component and this type of circuit component approximates closely to a constant current generator.

Examination of manufacturers' data for MOSFETs will show that quoted switching times are conventionally given for resistive load conditions. The unfortunate outcome of this state of affairs is that this information is almost useless to the circuit designer. Few if any power circuits encompass a purely resistive load. Even the ordinary magnetic loudspeaker represents a complex load. The result is that the only viable approach to predicting circuit performance is to calculate the switching losses. These calculated losses should always be confirmed until suitably good correlation can be achieved.

It is perfectly feasible to predict the total switching losses in a MGT by using equation (2.1), provided that certain additional items of information are also available to the designer.

By way of example the following study concept should be evaluated to gain an insight into the practicalities of the preceding statements.

A universal input off-line switching pre-regulator (Buck converter) with a maximum load current of 4 A is required to switch at a frequency of 70 kHz. Determine the maximum switching losses for the power switch which is given as an IRF840 for a gate input turn-on current of 200 mA. Assume that the turn-off gate current is 250 mA. Both values of gate current are considered to be average values. The reservoir (post-rectification) filter capacitance is given as 3300 μF.

GATE CONTROL

Given

(1) $V_{in[max]\,a.c.} = 264$ V (UK line nominal value 240 + 10%).
(2) $i_{g[on]} = 0.2$ A.
(3) $i_{g[off]} = 0.25$ A.
(4) $I_d = 4$ A.
(5) Switch frequency $f_o = 70\,000$.
(6) $C_{res} = 3300$ μF.

Initially assume:

(a) The filter inductor has minimal self-capacitance. (The self- or shunt-capacitance of an inductor is usually controlled by winding techniques.)
(b) V_f (the peak forward voltage) of the fly/free-wheel (commutation) diode is less than 0.3% of the d.c. bus and can therefore be neglected.
(c) 0.7 V $V_{f[rect]}$ (forward voltage) each limb for a conventional line rectifier bridge.

Therefore across C_{Res}:

$$V_{ripple} = I_d t_{ripple}/C_{Res}$$

$$= 4 \times 9.91/3300\text{E}^{-6}$$

$$= 12 \text{ V (after rounding down)}$$

$$V_{in[max]d.c.} = (V_{in[max]a.c.} \times \sqrt{2}) - (2V_{f[rect]})$$

$$= (264 \times 1.414) - 1.4$$

$$= 372 \text{ V (after rounding up)}$$

$$V_{ave\,d.c.[max\,value]} = V_{in[max]d.c.} - (V_{ripple}/2)$$

$$= 372 - 6$$

$$= 366 \text{ V}$$

From Figure 2.1 and using equation (2.1) the current rise time (t_{ri}) will be:

$$t_{ri} = Q_{tri}/i_{g[on]}$$

$$= 4 \text{ E}^{-9}/0.2$$

$$= 20 \text{ E}^{-9} \text{ s}$$

Similarly it can be shown that the voltage fall time (t_{fv}) will be:

$$t_{fv} = Q_{tfv}/i_{g[on]}$$

$$= 22 \text{ E}^{-9}/0.2$$

$$= 110 \text{ E}^{-9} \text{ s}$$

28 GATE CONTROL

The turn-on energy (e_{on}) will be:

$$e_{on} = V_{ave\,d.c.}I_d \times 0.5 \times (t_{ri} + t_{fv})$$
$$= 336 \times 4 \times 0.5\ 130\ E^{-9}$$
$$= \mathbf{95.16\ E^{-6}\ J}$$

Likewise the voltage rise time (t_{rv}) may be shown to be:

$$t_{rv} = Q_{trv}/i_{g[off]}$$
$$= 22\ E^{-9}/0.25$$
$$= 88\ E^{-9}\ s$$

Also, the current fall-time (t_{fi}) will be:

$$t_{fi} = Q_{tfi}/i_{g[off]}$$
$$= 4\ E^{-9}/0.25$$
$$= 16\ E^{-9}\ s$$

And the turn-off energy (e_{off}) will be:

$$e_{off} = V_{ave\,d.c.}I_d \times 0.5 \times (t_{fi} + t_{rv})$$
$$= 366 \times 4 \times 0.5 \times 104\ E^{-9}$$
$$= \mathbf{76.13\ E^{-6}\ J}$$

Finally the total energy per pulse (e_{tot}) may now be shown to be:

$$e_{tot} = e_{on} + e_{off}$$
$$= \mathbf{171.29\ E^{-6}\ J}$$

Therefore the total switching losses (P_{sw}) can now be calculated from:

$$P_{sw} = e_{tot}f_o$$
$$= 171.29\ E^{-6} \times 70\,000$$
$$= \mathbf{11.99\ W}$$

In order that the switching times may be deduced from gate-charge, by using the equation $Q = it$, the gate drive circuit of Figure 2.2 could well be used.

The gate input current for turn-on and the sink current out of the gate have both been set at 100 mA. The circuit of Figure 2.2 uses both halves of a dual high-current driver (SG3627) to drive a single MOSFET Q_1. Resistor R_2 sets the 'sourced' current into the gate of Q_1 and R_3 sets the sink current out of

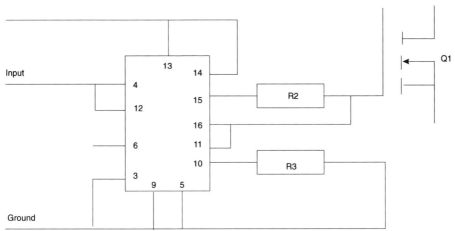

Figure 2.2 Gate drive circuit with constant source and sink currents

Q_1. The valves of R_2 and R_3 may be determined from:

$$R = 0.7/I_g$$

where the data sheet current sense voltage of the SG3627 is given as 700 mV typically.

There is nothing magical about the value of the current sense voltage. It is the typical value for V_{be} of a small-signal BJT. By varying the values of R_2 and R_3, thereby making i_{source} greater than i_{sink}, the turn-on speed may be made significantly higher than the turn-off, or conversely the reverse conditions could be made to apply.

The circuit of Figure 2.2 suffers from several major limitations, which may not become immediately apparent. One of the most serious of these drawbacks is the relatively high cost. Another is the poorly defined 'static' off-state condition of Q_1, since V_{gs} will not be reduced to a level much below the threshold value for Q_1.

A lower-cost alternative without constant current turn-on and turn-off would be to use a resistor in series with the gate. This has the effect of making the voltage rise and fall times to be equal and the same is also true for the rise and fall times of current.

It is relatively simple to contour the switching times to be unequal and simplified gate drive circuits to this effect are given in Figure 2.3.

In Figure 2.3(a) the turn-on speed will be greater than the turn-off speed if R_2 is made considerably smaller than R_1. The converse is true for the circuit of Figure 2.3(b).

GATE CONTROL

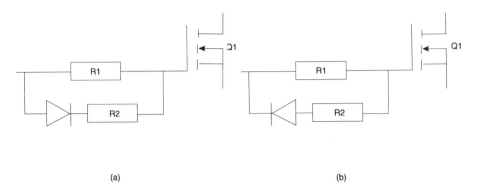

Figure 2.3 Simplified gate circuits for controlling switching times

The major points of interest in the two circuit variants being:

(1) The easing of the designer's task to contour the switching behaviour of the MGT merely by reversing the diode D_1 and thereby causing a reversal in the switching behaviour of Q_1.
(2) The accuracy achieved with either variant of the circuit leaves a lot to be desired. The prediction of times is not simply a matter of defining the relative time constants since strays (capacitance, resistance and inductance) can influence the behaviour in several ways. For the beginner the fact that the value of the two resistors being quantified empirically can give the basic circuit a certain degree of desirability.

However, the more advanced reader may wish to determine the values of the resistors by other means. This may be achieved by the use of first order approximations. It is useful in this case to know the supply voltage of the gate drive circuit which we will designate as v_{GG}. Since the drive circuit is driving an R–C load, where C is the input capacitance of the MGT, the total turn-on time may be predicted by the use of:

$$t_{on} = Q_G(Z_O + R_{G[on]})/(v_{GG} \times 0.47) \qquad (2.2)$$

where Z_O is the output impedance of the drive circuit and $R_{G[on]}$ is the gate series resistor during the turn-on phase.

Likewise the turn-off time behaviour can be predicted by the use of:

$$t_{off} = Q_G(Z_O + R_{G[off]})/(v_{G[clamp]} \times 0.43) \qquad (2.3)$$

where $R_{G[off]}$ is the gate series resistor during the turn-off phase and $v_{G[clamp]}$ is the clamped terminal gate voltage prior to turn-off (if a gate-source clamp Zener has been used). The circuits, of Figures 2.2 and 2.3, can be used individually and without recourse to any additional circuitry. Unfortunately, it is frequently desirable to provide some degree of galvanic isolation between the driven MOSFET and its associated control circuitry. The galvanic isolation may be the direct result of the need to comply with international safety specifications. The isolation can be achieved by a variety of methods. These may be:

(1) Isolation by drive transformer.
(2) Floating supply and opto-coupling of the input signal. (This latter technique enables the use of a *current sensing/(mirror) MOSFET* or *linear detection* to be employed in order that overload protection can be incorporated locally.)
(3) Power integrated circuits with in-built level shifting (but without true galvanic isolation).

Readers who are cognisant of the difficulties in using any one of the three techniques to drive BJTs will be satisfied with the degree of simplification which may be achieved with the implementation of any one of these techniques, when the chosen technique is applied to driving an isolated gate structure. This is especially true for transformer drive circuits where the dynamic range for duty cycle may well be large, namely where the ratio of $D_{(max)}$ to $D_{(min)}$ can be greater than a factor of 10. Difficulties of a different kind are also prevalent with transformer drive circuits, if the range of frequencies covered by the same drive circuit have a similar ratio for $f_{(max)}$ to $f_{(min)}$ as for the ratio for D.

Soft ferrite materials are severely lacking in the necessary volt-second saturation capability for operation at frequencies substantially lower than 10 kHz without the core becoming excessively large. (The figure of 10 kHz should be treated as a very crude approximation for the lower limit, since the variations which can be found in soft ferrites are so diverse that it is possible under certain circumstances to be able to work down to about 6 kHz.)

If it is desired to work down to a few tens of cycles per second, then an alternative core material would normally be a prerequisite, where the alternative core material will of necessity be significantly increased in bulk. The reader will be aware of the laminated cores used for the construction of line transformers.

The two circuits of Figure 2.4 are variations on the same theme and make use of the MOS structure's unique drive capability being somewhat similar to the drive capability of the thyristor. The similarity, between the two devices, is that once they have been turned on the devices will continue to conduct for a considerable period in time, provided that the gate voltage of the MOS-gated device remains at the full enhancement potential.

The input signal for both variants of Figure 2.4 should be supplied via a totem-pole driver to the same input terminal, namely, Z_1.

The two circuits of Figure 2.4 may be used in circuits without modification for frequencies as diverse as 1 Hz to 1 MHz and for duty cycle variations from approximately 1% to 98% without suffering any penalty in drop of the output waveform.

32 GATE CONTROL

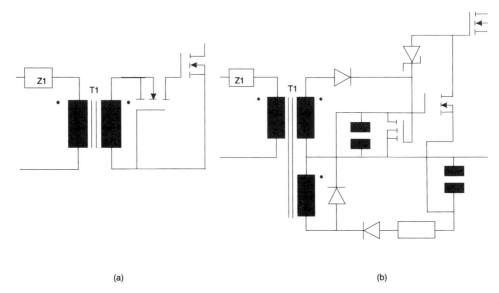

(a)   (b)

Figure 2.4 Transformer isolated gate-drive circuits for low frequency and/or wide duty-cycle variation usage (Courtesy of International Rectifier Corporation)

Both of the circuits behave in a manner that is fundamentally the same. This behaviour can best be summarised as follows:

With a positive signal at either input the full signal voltage is applied across the primary of T_1. Current flows within the parasitic diode of the small driver MOSFET (in (a)) and through a discrete diode and Zener (in (b)) and charges the gate of the driven MGT until it turns fully on.

After a finite interval in time the core of T_1 will saturate and the voltage at the secondary of T_1 will collapse to zero. The driver MOSFETs parasitic diode (discrete in (b)) is reverse-biased and the gate of the driven MOSFET is virtually left in a state of limbo. Current through the transformer primary is limited by Z_1 to a safe value. Z_1 may be any type of reactive component; but the optimum results will arise out of Z_1 being a capacitor. Provided that its gate-voltage is not seriously altered, the driven MGT will continue to conduct as if no change has taken place at the secondary of T_1.

When the signal is removed from across the primary of T_1, the transformer core will recover and the voltage across the secondary will reverse to an opposite polarity to that of the original signal which existed prior to saturation. This negative voltage at the secondary of T_1 will cause the small driver MOSFET, whose drain is connected to that of the much larger (driven) MGT, to discharge the input capacitance of the driven device and so turn it off.

GATE CONTROL

There are obvious differences in the two circuits of Figure 2.4 and these occur because of the conflicting requirements placed upon them. Figure 2.4(a) is meant to be a no-nonsense low-cost solution without frills, whereas the circuit of Figure 2.4(b) is meant to provide a degree of reverse bias to the gate of the driven MGT (for the duration of its off-state). The circuit of Figure 2.4(b) is better suited to applications where high values of dv/dt may be applied across the driven component during the period when it is meant to remain 'off'.

Selection of the magnetic core for the transformer T_1 is simplicity itself. If used correctly the following rules will be found to provide all of the necessary guidance to make a satisfactory design of a fully working circuit.

(1) Determine what will be the maximum voltage which is likely to be applied across the primary of T1 (this may well be the supply voltage of the control circuitry on the primary side), and assign this voltage to the variable V_{ss}. Make $V_{gh} = 10$ V.

(2) Determine the maximum value of C_{iss} for the driven device. If manufacturer's data is only supplied for a typical value then allow for a further 10% above this typical value.

(3) Make Z_1 sufficient to limit the primary current of T_1 to be commensurate with the totem-pole driver's capability, on the input side. Ideally, Z_1 should be a capacitor (C_{in}) and resistor (R_{in}) connected in series.

(4) Make $C_{in} > 2C_{iss}$ (from (2) above).

(5) Time constant t_p will by definition be:

$$t_p = 0.67 C_{in} R_{in} \tag{2.4}$$

(6) Make β_{max}, for the core, approximately equal to 80% of the value of β_{sat} (from core data) for a core temperature of 100 °C. This value for β_{max} should be more than sufficient to guarantee core saturation.

(7) Make β_{op} the operating flux density for the core of T_1, where:

$$\beta_{op} = 0.7 \beta_{max} \tag{2.5}$$

(due allowance has been made for $\beta_{remanance}$).

(8) Obtain usable core area A_e from core data to satisfy the condition where:

$$A_e = (V_{gh} t_p)/(N_p \beta_{op}) \tag{2.6}$$

where N_p is the primary turns and should ideally be set between 10 and 15.

(9) Calculate the secondary turns N_s from:

$$N_s = V_{gh}/(V_{ss} N_p) \tag{2.7}$$

(10) If the value of N_s is not an integer, round up the value for N_s to the next higher integer value and use a Zener clamp on the gate of the driven MGT. Repeat steps (9) and (10) if there is more than one secondary

winding, and where each winding may drive MGTs with differing gate structures.
(11) Scale the diameter of all the winding conductors to have a current density of less than 3000 A/in^2 (4.65 A/mm^2) of cross-sectional area for commercial applications and for a current density of less than 2000 A/in^2 (3.1 A/mm^2) for military applications.

The design of the saturating core transformer, for both versions of transformer in Figure 2.4, may be regarded as being complete. If the design procedure, outlined above, should at first appear to be too tedious, all that can be said is that a little practice will soon dispel this notion.

The saturating core transformer imposes a major dilemma upon most people who are accustomed to designing transformers. This dilemma arises out of the need to overcome an in-built aversion to core saturation. Magnetic core saturation is normally considered to be one of the major taboos to the magnetic specialist. The sole exception to this fear of core saturation is fundamental to the design of the blocking oscillator (and other transformer coupled self-oscillatory circuits). It is also one of the fundamentals of design associated with magnetic amplifiers. On this occasion the advice is that the core is required to saturate, for the practicality of reducing the size of the transformer.

If the volt-second saturation capability has been correctly selected then a duty ratio of less than 10% at the maximum frequency of operation should be eminently possible, whilst allowing a duty ratio of greater than 90% at the minimum frequency.

The reader may already have a preference for using some form of floating supply and an opto-coupler. Power BJT users will be only too aware of the problems associated with the implementation of a floating supply and opto-coupler driver input. The critical parameter which poses the major stumbling block for this particular technique with BJTs is the absolute power level of the floating supply. The low drive power level of the MGT can overcome this stumbling block both neatly and elegantly. If the required frequency should not exceed 1 or 2 kHz then an extremely novel component may be considered as a useful alternative to the floating supply and opto-coupler combination. The device in question has the title of Photo-Voltaic Isolator, which has been generally abbreviated to the acronym — PVI.

The PVI has the unusual ability to supply an output voltage of 5 or 10 V, depending upon device type, at an output current of approximately 10–20 μA simultaneously, when the input Light Emitting Diode has been illuminated. The absolute maximum output current from the isolator is approximately 50 μA, under current-limit operation. This limited output current capability of the PVI is of little consequence to this application, owing to the very limited drive power required by the MGT, especially at the modest frequencies for which this component is eminently suitable.

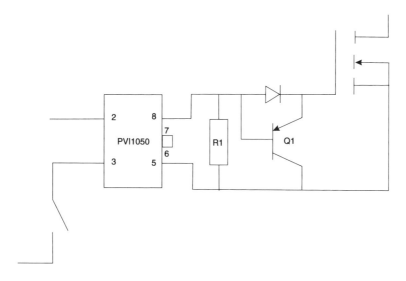

Figure 2.5 Simple gate-drive circuit with PVI

The circuit of Figure 2.5 depicts a logic-level MOSFET being driven by the PVI.

It should be noted that no special circuitry for protecting the MOSFET has been included. The reader is encouraged to use their imagination and ingenuity to devise any (or all) of the plethora of protection circuits if it is considered to be either desirable or even necessary.

If the desire to include some form of protection is deemed as being necessary, then I would advise the use of any one of the many micropower or CMOS linear integrated circuits which are available at present. The need for these types of IC should be self-evident, being warranted by the limited current capacity of the PVI. The turn-off PNP should be a very high gain Darlington and the diode should be a *low leakage type*.

Use of the PVI makes possible the rapid design of extremely simple gate drive circuits.

The only additional component that is required is a moderately low impedance turn-off circuit, to prevent accidental turn-on by applied dv/dt. This circuit also fulfils the additional function of being the turn-off function. Without this additional circuit turn-off of the PVI driven MOSFET might never occur.

An alternative circuit, with the inclusion of current-limit circuit components, is depicted in Figure 2.6. Current limiting is performed by the circuit which includes R2 and the small-signal NPN Q2.

The procedure for selecting the value of R2 will be given in a later chapter. Here the inclusion of two PVIs has been included owing to the present non-availability of logic-level current-sensing MOSFETs and the desire to possibly use one of these devices. A single 10 V PVI could be used if it was so desired. Please note that the components in Figure 2.6 are freely available. I

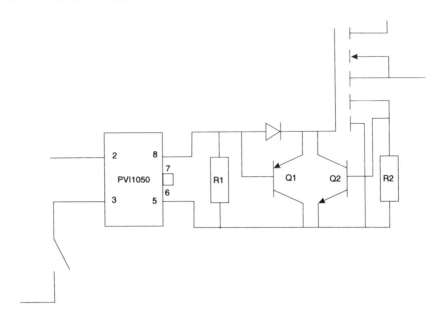

Figure 2.6 Simple gate-drive circuit with PVI and current-sensing MOSFET

would again draw attention to the fact that several alternative ideas may easily be devised. If increased complexity is required as a result of additional functions being included then the use of micropower ICs is again recommended.

It should be noted that in both of the circuits of Figures 2.5 and 2.6 the inclusion of the three components of D1, Q1 and R1 are virtually unavoidable, in order that the turn-off circuit be defined. Because of the limited current capacity of the PVI the value of R1 should be such that between 1 and 10 μA is allowed to flow through it. The implied minimum value of R1 would therefore be about 500 kΩ. It should also be remembered that the value of R2 will decide the turn-off speed of the 'MirrorFet' Q3. The setting of turn-off speed by the value of R1 will be discussed in greater detail in a later chapter.

The two circuits of Figures 2.5 and 2.6 may be used to configure either 'high-side' or 'low-side' switches with an equal degree of confidence. The terminology of high-side and low-side will be discussed more fully in Chapter 10.

One aspect which requires stressing is that the circuits of Figures 2.5 and 2.6 can be used, with absolute confidence, in conditions where the MOSFET, or other MGT functions, is utilised as a *static* switch.

It is also possible to use monolithic integrated circuit technology to drive high-side and low-side switches which are *bridge* or totem-pole configured or connected in a 'two-transistor' configuration. The monolithic circuits which are presently available can drive totem-pole configured power switches which operate in converter circuits with supply voltages rated at up to 500 V d.c.

GATE CONTROL

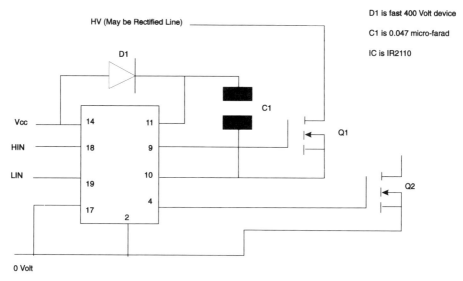

Figure 2.7 Isolated gate drive circuit using a bridge driver IC

One of the MOS-gate drive ICs is capable of driving both the high-side and low-side power switches of the two-transistor converter configuration. This particular circuit configuration is also used in motor drives which are used to control reluctance motors.

This supply off-set capability makes them ideally suited for off-line 220 V operation where the maximum value of the rectified line will be 372 V.

One such IC and its application is illustrated in Figure 2.7.

The availability of this type of IC is not confined to a single supplier. In the circuit in Figure 2.7 the centre connection of the totem-pole has been intentionally left *open*, since this particular IC has been provided with two independent channels, and can therefore be used in the type of power converter topology on occasion referred to as an asymmetric half-bridge; the nomenclature is found to be quite erroneous in this instance. Even though the use of this type of IC is meant to greatly simplify the design of high-voltage high-side switch driver circuitry, it is none the less imperative that the reader should be aware of the need to maintain strict observance of the recommendations relating to the good practices of layout and decoupling.

The use of *current pulse mode level shifting* has reduced the measure of common-mode rejection that is normally associated with integrated circuit technology which is traditionally utilised for driving totem-pole power stages. This reduction in common-mode rejection can have certain serious consequences, if its connotations are neglected. In bridge connected totem-pole power circuits a finite dv/dt will always be observed across either one of the power switches, during switching transitions. The applied dv/dt across the high-side switch — and it is not important whether this switch is 'off' or 'on' — if in excess of the

38 GATE CONTROL

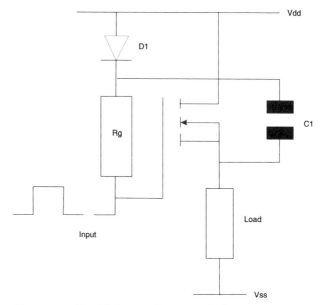

Figure 2.8 Simplified circuit illustrating the bootstrap concept

maximum rated value of applied dv/dt to the upper channel, can result in false triggering of the upper switch. The maximum value of applied dv/dt for the upper channel is a direct consequence of the limited common-mode rejection of this particular channel, due to the type of level-shifting which has been integrated. This last paragraph is not meant to denigrate the usefulness of this type of IC. It is meant to advise the reader into maintaining strict adherence to ratings and the observance of good circuit practices. The data sheet value for dv/dt is sufficiently high to ensure that it will seldom, if ever, be approached in practice. This cautionary note is given for the benefit of experienced users who may not be conversant with IC gate drivers.

In the circuit of Figure 2.7 the high-side channel of the gate driver IC is supplied with a type of semi-floating power source. This type of power source utilises a *bootstrap capacitor.*

The term 'bootstrap' has been derived from the notion of the physical analogy of humans being able to bend down and lift themselves up into the air by their bootstraps, thus enabling the human being to 'fly'. The electrical equivalent to the physical analogy does not involve such 'flights of fancy'. In the circuit of Figure 2.8, the source/emitter of the MGT, upon being switched on by the positive going pulse at the input end of R_g, will move in potential to be virtually the same as the drain or collector. The change in voltage at the source/emitter is transferred to the upper plate of the bootstrap capacitor C_1. This change in voltage at the cathode of D_1 causes the diode to be reverse-biased and to turn off. The cathode of D_1 can now be raised by the full voltage which was across the MGT in its off-state. The supply to the gate pull-up resistor R_1 is, therefore, seen to change by the same value in voltage as the gate needs to be raised as the original potential drop across the MGT. This change in supply voltage to the gate pull-up resistor enables the MGT source/emitter to *follow* the gate.

An empirical formula for the scaling of C_1 is given as:

$$C_1 \geq C_{iss(Q1)} \times 10 \tag{2.8}$$

Diode D_1 should be a low-powered component with a reverse voltage capability equal to the supply voltage V_{cc} (at least). The reverse recovery time of D_1 should be commensurate with the turn-on speed of Q_1.

Bootstrapping can therefore be seen to provide a convenient technique for providing 'high-side' drive energy inexpensively.

The 'trade off' which must be made for this convenience is in 'turn-on' speed. The reduction in 'turn-on' speed is seen to become evermore pronounced as the supply voltage to Q_1 is increased.

A further factor affecting the turn-on time of Q_1 is the value of the gate pull-up resistor R_1. This component should be scaled sufficiently to enable the current through the gate protection Zener D_2 to be optimised for dissipation in both R_1 and D_2. Replacing R_1 with a small-signal transistor to provide active pull-up would nullify the speed reduction, but would do so at the expense of increased cost.

It is also perfectly feasible to replace Q_1 with a device which is the complement of the original component, especially if the original part was a MOSFET, i.e. replace the N-channel device for a P-channel part. The drain and source connections would have to be reversed and bootstrapping would be found to be unnecessary. The polarity of the gate signal would also have to be reversed. This apparent further simplification is unfortunately only suitable for P-channel switches with a break-down voltage equal to or less than 200 V. Higher voltage parts tend to have on-state losses which are unacceptable, especially if these parts are MOSFETs.

The complement of the N-channel derived IGT would be prohibitively expensive for virtually all applications and the savings which could be accrued from bootstrapping would be negated by the cost of the component. Since the P-channel derived IGT would tend to be used in a 'cost no object' application the added cost of more efficient drive circuits would be proven to be worthwhile.

In the design of gate-drive circuits, it is recommended that one other golden rule be observed at all times: *the impedance of the turn-off network should be kept as low as possible*. This measure has the tendency of increasing turn-off speed, which may prove to be embarrassing, but provides the essential benefit of reducing the likelihood of inadvertent 'Miller turn-on'.

Extremely high values of applied dv/dt could cause Miller turn-on even though the recommended precautions of maintaining low turn-off impedance have been

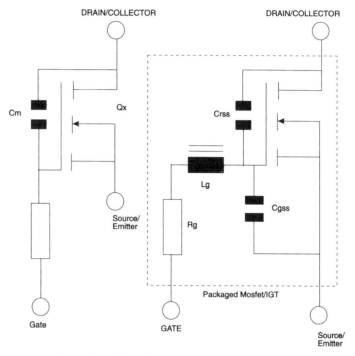

Figure 2.9 Equivalent circuit of power MGT gate

observed. This unfortunately arises out of several factors such as gate bulk resistance and/or the effect of package strays.

The effect of package strays tends to be exacerbated by the traditional metal packages (TO-3, TO-66 and TO-39) so popular with the military. These packages invariably use a steel flange as the base of the package and the effect of this ferrous flange is to considerably increase lead inductance, especially the leads passing out of the package through a glass to metal seal.

The effect of the flange is to behave as a magnetic core. The effect of the glass to metal seals is considerably reduced in the typical 'plastic' type of packages.

An explanation of why the style of packaging of the MGT may be influential in Miller or dv/dt turn-on is given in Figure 2.9. The left hand side of Figure 2.9 shows the equivalent circuit of the gate of the MGT, whereas the right hand side illustrates the gate associated strays of the packaged part. If the equivalent circuit in Figure 2.9 is considered, it will become obvious to the reader that the combined capacitance C_m is the sum of gate to source C_{gss} and reverse transfer or Miller C_{rss} capacitances. This capacitance C_m, along with the lumped impedance Z_g of the gate-drive circuit, forms a dv/dt potentiometer. If any value of dv/dt is applied across Q_X, a current i_m will flow through C_m. The magnitude of this displacement current will in turn develop a voltage across Z_g. If the magnitude of this potential, across Z_g, happens to exceed the threshold voltage of the MOSFET, then turn-on of Q_x will be inevitable.

The action described in the preceding paragraph is frequently referred to as Miller turn-on. The phenomenon of Miller turn-on has been alluded to by other authors under several diverse titles. The variety of titles, under which this subject has been categorised, can easily lead to confusion. I will therefore present some of the more common headings in order to enlighten the first-time user, and therefore reduce the possibility of confusion. Some of these alternative titles are:

(a) static dv/dt turn-on,
(b) Mode 1 turn-on,
(c) noise turn-on.

2.1 CURRENT LIMITING AND VOLTAGE CLAMPING BY GATE CONTROL

It is a well known fact that both current limiting and collector voltage clamping have been frequently used with BJTs. Almost identical techniques can be used for MGTs with equally satisfactory results. Since both of these techniques form part of subsequent chapters, they will therefore not be covered in this chapter. The reason for their introduction has merely been to enlighten the reader as to their potential usage.

The final recommendations which are advocated within this chapter relate to the very real need to protect the integrity of the gate oxide of both power MOSFETs and IGTs. If these devices exhibit any fundamental weakness, then it has to be connected with the relative fragility of the gate structure. This structure and the consequential possible failure mechanisms were covered more fully in the preceding chapter and should therefore need no reiteration. It therefore becomes necessary here to define some of the techniques which may be adopted as the means of ensuring that the integrity of the gate is not violated, if absolute circuit reliability is to be maintained.

In-circuit protection is easily achieved by the judicious use of relatively low-powered Zener diodes, connected from the gate to the source of the device requiring protection. When using gate protection Zeners it is vital that the following rules be strictly applied.

(1) Situate the Zener and its connections as close as possible to the actual gate and source terminals of the device which the Zener is meant to protect. Remember that 1 cm of wire or printed-circuit board track equates roughly to 10 nH of inductance.
(2) Keep the wattage, and therefore the size, of the clamp Zener as low as possible. This ensures optimisation of the clamping speed of the Zener, since

the speed of clamping is directly related to the junction capacitance of the zener. The implication of this last statement is that the smaller the junction, then the lower will be the junction capacitance and therefore the time to clamp any spike which may occur.
(3) Always ensure that the gate-drive circuit is provided with the correct degree of decoupling. This correctness extends both to value, quality and the siting of the decoupling capacitor. *It is false economy to skimp on the quality of the decoupling capacitor.*
(4) Always site the decoupling capacitor as close as possible to the actual gate-drive circuit. *Never allow PCB aesthetics to dictate the positioning of this vital component.*

Handling precautions have been referred to in the preceding chapter. The precautions cited were mainly concerned with the environment which could be encountered within development laboratories. A few words on the subject would not be amiss at this juncture, but the discussion will now be moved to alternative locations.

During manufacture of printed-circuit boards and equipment it is equally important that electro-static damage precautions continue to be observed. These precautions should be maintained both at the point of storage and along the entire path, right to the point of despatch of finished assemblies.

It is prudent to reiterate any rules which may have been stated previously.

(1) Avoid unnecessary handling of individual components. If the handling of the part is deemed to be unavoidable, then the use of *ground straps* should be regarded as being mandatory.
(2) All floors and working surfaces should be coated with a suitable *anti-static* coating, and it is advisable that *anti-static mats* be placed upon the working surfaces.
(3) The effectiveness of the anti-static coatings and mats be periodically checked. It would be unfortunate if these protective measures were to be installed and forgotten.
(4) Operatives working with *static-sensitive* components should be constantly reminded of their obligations to observe all ESD prevention procedures.

3
Over-voltage Protection

Fully protected MOSFET switches have recently been launched by some manufacturers. Although this technology may be regarded as still being in its infancy it is a certainty that other suppliers will follow in the wake of the original vendors. The reader should understand that not all devices will be fabricated using this monolithic fabrication process. A large number of standard devices will remain on the open market. These devices will require their environment to be made as conducive, as possible, to ensure their reliability. Therefore their protection will be required. The next two chapters will be devoted to the subject of the MGT's protection.

In the preceding chapters over-voltage protection was referred to in a general sense, and any specifics of the subject were mainly related to over-voltage of the gate. Unfortunately, over-voltage within any circuit will not necessarily be confined to any one part of that particular circuit. It can and usually does occur anywhere. If this were not enough, the series over-voltage will be found to occur at the least opportune moment when it can easily cause the most embarrassment. The real cause of failure will be found to occur outside of the development environment, which may be the development laboratory, where mains supplies tend to be between specified limits and relatively clean. Out in the field, where the equipment will normally have to operate, conditions may well be grossly inferior and these conditions should be catered for in the overall design.

The text of this chapter will be confined to the major power lines within the power part of the circuit. Any over-voltage on these power lines will tend to apply a severe stress mainly to the drain-source/collector-emitter regions of the power MGT. It should be emphasised that the over-voltage transient, if sufficiently rapid with respect to the leading edge of the transient, can be coupled through the 'Miller' capacitance of the MGT on to the gate. If gate protection, as recommended in Chapter 2, has been rigorously applied, then further consideration for the gate (as far as over-voltage stresses are concerned) will be unnecessary. However, over-voltage protection of the gate is the most frequent failure mode of MOS-gated devices and is the easiest failure mode to protect against. It is therefore one of the subjects that I will repeatedly emphasise.

Experienced users of BJTs will be only too familiar with the consequences of exceeding the collector to emitter voltage capability of that particular device. They will be only too aware of the dangers associated with *second-breakdown*. Second-breakdown within the BJT unfortunately is not merely associated with excessive voltage stress. It also tends to have significant temperature dependency and may also be associated with the drive circuit and the way the device is turned on/off.

Drive circuits (if poorly designed for BJTs) may well have the effect of causing 'current crowding/current hogging' to occur within the dice resulting in localised 'hot spots' which in turn may lead to second-breakdown. Since this book is not overly concerned with the niceties, or otherwise, associated with BJTs, and driving them, no further effort will be expended on this particular subject.

It is fortunate indeed that the true power MOSFET structure is not affected by the undesirable phenomenon of second-breakdown. The IGT, being a derivative of the MOSFET, is also blessed to some extent with this immunity and may be evidenced in data sheet limits of the overload current capability of the device. Unfortunately, the same cannot be said to apply to the parasitic transistor (if the device is a MOSFET or the parasitic thyristor of the IGT), within the MOS structure. If emitter injection of the parasitic transistor can be induced then second-breakdown or latching of the parasitic will be the inevitable result. The need to protect against this eventuality is the objective of this chapter.

Strict observation of the details proposed in this chapter can be neglected to some extent if, during the design phase, certain precautions are taken to reducing stray and other forms of inductance as much as possible. This applies particularly to the leakage inductance of power transformers. The spikes generated by this component must, in nearly all cases, be limited for safety, if for no other reason. Would it not be more desirable to reduce the spikes at source, by careful design of the transformer?

It is accepted that the best possible design of transformer may unfortunately retain a high value of leakage inductance. The point being made is that wherever possible the reader should strive to reduce the value of this highly undesirable element within the transformer.

If, on the other hand, component design cannot be relied on for reduction of transients then other solutions become mandatory.

The various techniques which may be adopted are discussed at some length, and the incorporation of some, if not all, of the precautionary measures may have to be invoked. The various forms of protection against over-voltage will be presented in alphabetic order and not from any sense of personal preference.

3.1 ACTIVE CLAMPS

During the switching excursions of the power MGT and it should be remembered that these excursions can be extremely rapid and brief in the case of MOSFETs

3.1 ACTIVE CLAMPS

where rates of change of current (di/dt) of 10 000 A per μs are not unheard of, the stray circuit inductances will by their very nature generate unwanted transient voltages which may well exceed the breakdown voltage capability of the device which is performing the switching action. The magnitude of these transients may be deduced from the following:

$$e_{\text{transient}} = \int L_{\text{STRAY}}\, di/dt \qquad (3.1)$$

For values of i which may be a few amperes at most and where t may be a few tens of nanoseconds equation (3.1) is adequate since the effects of conductor resistance may be neglected.

Where i may well be a few tens of amperes and t may be measured in hundreds of nanoseconds we find that R_{circuit} becomes significant and we need to redefine $e_{\text{transient}}$. The peak value of $e_{\text{transient}}$ may be found using:

$$e_{\text{transient}} = \frac{i_{\max}[(2L_{\text{stray}}) - (tR_{\text{circuit}})]}{2t} \qquad (3.2)$$

where $i_{\max}$ is the peak–peak value of the current excursion and t is the duration of the excursion of $i_{\max}$. R_{circuit} is the sum of the resistances to be found in circuit and may include the resistance of any stray inductances if this resistance is significant.

If the MGT in question is unable to withstand the transient over-voltage (without the parasitic BJT transiting into second-breakdown, or the parasitic thyristor from latching) then some form of voltage clamping around the MGT must be invoked. One type of voltage clamp is a transient suppressor diode but, because this device will be described in its own right, it will not be discussed any further, for the present. A second form of voltage clamp is to utilise the MGT itself as an amplifier, in conjunction with a Zener diode, to magnify the clamping action of the Zener diode. This type of circuit has been frequently used with BJTs especially when these transistors have been used in automotive ignition systems. The active clamp of the inductive discharge ignition circuit maintained the switching trajectory of the ignition Darlington within its turn-off safe operating area.

One form of active clamp is demonstrated in the circuit of Figure 3.1. The action of this circuit can best be described as follows. Let us consider what occurs over the duration of a transient waveform of voltage which may be applied across the MOSFET Q_x from drain–source. As the voltage at the cathode of D_z is increased, either progressively or rapidly, D_z itself will start to turn-on and pass current to the gate of Q_x.

When Zener diode D_z has been fully turned-on, the drain and gate of Q_x may well be considered as being inexorably linked.

Progressively increasing the voltage at the drain of Q_x will cause both C_{GS} and C_{RSS} to be charged and result at some time in the threshold value of the V_{gs} for Q_x being exceeded, and consequently lead to the turn-on of Q_x.

46 OVER-VOLTAGE PROTECTION

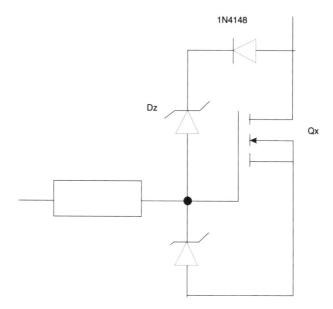

Figure 3.1 Simplified circuit of active clamp

Any further tendency for the voltage at the drain of Q_x to increase will only lead in turn to a further increase in the 'turning-on' of the MGT. This will in turn apply a clamp action to the voltage at the drain/collector terminal. Two further points need explaining.

The power capability of the Zener of necessity need be no greater than 100 mW. The 1N4148 diode in series with the Zener is to prevent 'Baker clamp' action from taking place. The term 'Baker clamp' is a carry-over of anti-saturation techniques from BJT designs and is only used to clarify a point; it will therefore not be necessary to go any further into any description for the Baker clamp.

The circuit of Figure 3.1 will be found to function perfectly. Nevertheless a few words of caution are felt to be appropriate. Although the MGT is working within its active region, and if it is never allowed to stray outside of its Safe Operating Area, then the device will be found to be perfectly protected, provided the dissipation does not prove to be excessive for any heatsinking or for the dissipation capability of the device itself.

3.2 AVALANCHE

Although the MOSFET structure's drain–source break-down mechanism is fundamentally that of avalanche, the reader is cautioned to be aware that MOSFETS bearing the same part number, but from differing vendors, may well not share any common characteristics. Some, as opposed to all, MOSFETs are endowed with a commensurately greater degree of ruggedness over those from other manufacturers. This ruggedness takes the form of being able to absorb a significantly greater level of energy, in avalanche, than the lesser types. The ability to absorb high levels of energy whilst in avalanche, or to withstand the flow of high values of current when being subjected to the stress of avalanche breakdown, may on occasion be the result of the good fortune of the manufacturer of the MOSFET. Normally this form of ruggedness will have been designed-in by the manufacturer. It is of no concern how the particular MOSFET of choice came to be provided with its level of ruggedness. It is sufficient to be aware of the device's capabilities as specified in the data sheet and to be able to use it.

The IGT does not have anywhere near the same level of ruggedness in terms of avalanche capability as the MOSFET and will, therefore, not have this form of self-protection. It is therefore only the MOSFET to which this section will apply.

Avalanche self-protection is perfectly feasible provided the following rules are obeyed.

(1) If the over-voltage is an infrequent occurrence then the single pulse energy limit (E_{as}) must be observed. The value of E_{as} is found to be dependent upon the junction temperature, and this dependency should be kept in mind when using avalanche self-protection.
(2) On the other hand, if the transient is observed to be repetitive by nature, then the repetitive energy limit E_{ar} must be observed. This energy limit can be used to the fullest extent provided the overall power handling capability of the MOSFET is not exceeded.
(3) Whilst observing the limits of (1) and (2) above it should also be firmly borne in mind that the avalanche current rating I_{ar} should never be exceeded. It will be noticed that the maximum value of I_{ar} is absolute and applies with equal validity to all avalanche conditions, whether these conditions result from a single pulse transient or alternatively if the transient is of a repetitive nature and created by the occurrence of a train of pulses.

The outcome of being able to use the avalanche capability is the effect this can have on making the power circuit cost effective, since repetitive avalanche operation of the MOSFET may save the cost of secondary voltage clamping sub-circuits.

3.3 PASSIVE CLAMPS

Current snubbing

If the MGT being considered does not have an avalanche capability, and/or the user is unwilling to use active clamping then the use of a *current snubber* also referred to by some people as *'slow rise circuits'* may well be considered as being necessary to maintain the turn-off locus within the SOA of the transistor. BJT users will no doubt be accustomed to the use of snubbers since these circuits were essential in preventing the BJT from straying into second-breakdown during turn-off.

The current snubbers referred to in this chapter are *turn-off snubbers.* These snubbers are usually one of two types. They are either of the *dissipative* variety or of the *non-dissipative* sort. There are several fundamental differences between the two types. These differences are:

(1) The dissipative snubber is certainly more simple to design, whilst at the same time being obviously less efficient than the non-dissipative variety.
(2) The component count of the dissipative snubber is very much smaller than the component count for the non-dissipative version.

The final choice of the type of snubber will therefore be made on the back of other variables such as cost, etc. The reader is advised to study both varieties carefully and to use this knowledge for each particular application.

Never abandon the possibility of using either type (in favour of the other), by prematurely discarding the variety which at first appears to be less cost-effective or more difficult to configure. Always decide on merit.

When designing the dissipative snubber the reader should be aware that there are two distinct versions.

The two types most commonly used are the *hard snubber* as in Figure 3.2(a) and the *soft snubber* of Figure 3.2(b). The reader should not confuse the hard snubber with the 'spongy clamp' — whose configuration and action is in many respects similar to the hard snubber, but still having sufficient variances to warrant the difference in classification.

The first-time user should become familiar with the fact that the two versions are intended for use in differing circumstances, the hard snubber being more suited to being applied to the single-transistor Buck converter circuit and to the transformer derivative of this type of circuit. It is also particularly suited for use in the Buck-boost circuit and its transformer derivative — the fly-back converter. The soft snubber is more suited to all variants of push–pull power circuits and is the only one that can be used with complete confidence in totem-pole power stages. There is a divergence from this philosophy, but the type of semiconductor concerned (being quite different) ensures that the validity of the rather loosely defined rule of totem-pole power stages is not brought into question.

It has been my experience to observe, under normal circumstances, that snubbers are not designed. They merely happen to evolve by empirical methods, or come into existence by the simple expediency of trial and error.

The seasoned designer of BJT power stages, who happens to be completely experienced in correctly designing snubbers, will be fully cognisant of the need

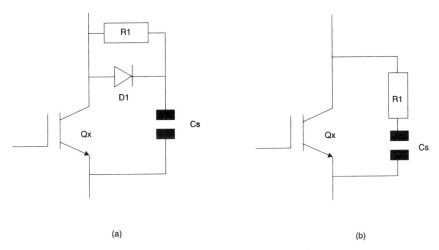

Figure 3.2 Hard and soft dissipative snubbers

to take into account the maximum current fall-time of the bipolar, during the design of the snubber. The necessity for this approach was the importance of maintaining the switching trajectory, to ensure that it always resided within the SOA of the transistor. The basic equation from which all calculations were derived was:

$$\hat{\imath} = c \, dv/dt \qquad (3.3)$$

where $\hat{\imath}$ is the peak value of the displacement current through the capacitance c for a given rate of change of voltage (dv/dt).

For use in snubber and clamp circuits we need to know the total excursion of voltage and we therefore need to know the charge which is transferred to the capacitor. This is given as:

$$\delta V = I_c \delta T / C_s \qquad (3.4)$$

where δV is the difference between $BV_{ceo(sus)}$ and $V_{ce(sat)}$, or reduced to its simplest form δV can be equated to $BV_{ceo(sus)}$. Likewise I_c corresponds to the maximum value of collector current, while δT is the maximum value of t_{fi} (the current fall-time).

Because of the extreme difficulty in accurately determining t_{fi}, it was quite normal for a compromise value or best guess estimate to be inserted into the equation. The nett result was that if the calculated snubber capacitor C_s was deemed to be too small for practical purposes, whilst ensuring that all conditions of $t_{fi(max)}$ were catered for, then the vast majority of transistors ended up being *over-snubbed for safety*. Even worse is the situation of transistors being *under-snubbed in the search for efficiency* while not making allowances for the worst case of t_{fi}.

Consider the same snubber (with all things being equal), but with the BJT now being replaced by a MGT. The technique used for calculating the value of C_s is vastly simplified. The only requirement is to know the energy stored in the circuit inductance in its entirety. This energy could be written as:

$$E_s = 0.5\, L_c \hat{i}_d^2 \tag{3.5}$$

where L_c is the sum of all the circuit inductances within the current path of the MGT, including stray, leakage and magnetising inductances of the transformer, if one assumes that the transformer snubbing is included within the MGT's snubber. Finally $\hat{i}_d$ is the peak value of drain/collector current. (The magnetising inductance of the transformer must be included if the magnetising current has an ill-defined shape and/or recovery path.)

Equation (3.5) illustrates that it is only the energy to be contained that must be dumped into the snubber capacitor C_s. This leads to the next equation:

$$E_s = 0.5\, C_s BV_{DSS}^2 \tag{3.6}$$

where C_s is the snubber capacitance and BV_{DSS} is breakdown voltage of the MGT.

Equating the two sides of E_s one gets the final MGT snubber equation:

$$L_c \hat{i}_{dc}^2 = C_s BV_{DSS}^2 \tag{3.7}$$

That is all that is required for calculating the snubber capacitance for the MGT. It is obvious that uncertainties relating to switching times are not included into the calculations. This is in complete contrast to the BJT snubber capacitor calculations. The value of the snubber resistor R_s bears a remarkable degree of similarity for the two types of power switch. The slight variation is the allowance which must be made for the 'dump current'. It should be borne in mind that the BJTs peak current capability is modest when it is compared with that of the MGT. The power dissipation capability of the discharge can be found by using:

$$P_r = C_s V_p^2 f / 2 \tag{3.8}$$

where P_r is the power dissipated by the snubber resistor, C_s is the snubber capacitor value (as calculated before), and V_p is the breakdown voltage of the switch, whichever type may be in use.

All of the techniques so far used for snubber calculations are for the hard snubber variation. Due allowance for the soft snubber must be made for the voltage across the capacitor and resistor which occur simultaneously. It is usual to halve the values for the voltage component, as used in the foregoing equations, to calculate both the value of resistor and capacitor. Because the capacitor is charged and discharged through the resistor, the equation for calculating the power dissipation in the resistor must be modified by deleting the denominator.

3.3 PASSIVE CLAMPS

The modifications of snubber rules of the previous paragraph concerning the halving of the value of the supply voltage should be applied with equal validity to converters of bridge and half-bridge topologies.

When designing non-dissipative snubbers, the reader should bear in mind that the 'normally regarded practice' is to return the energy — which has been dumped into the snubber capacitor — to the converter's supply. This immediately infers that non-dissipative snubbers lead to an overall improvement in efficiency and this proves to be the case.

The technique that is commonly used for the evolution of a non-dissipative snubber is to use a resonant circuit to discharge the snubber capacitor C_s into an auxiliary capacitor C_a, during the time the switch being snubbed is on. The resonant charge/discharge technique ensures the total discharge of C_s. It is necessary to ensure that C_a discharges during the time the switch is 'off'.

Given that:

$$\hat{i} = V_{supply}/(Z \sin wt) \tag{3.9}$$

and

$$V_{Cs} = V_{supply}[1 - (1/(1 + c_r))(1 - \cos wt)] \tag{3.10}$$

where $\hat{i}$ is the resonant current and V_{Cs} is the instantaneous voltage of the snubber capacitor during discharge.

$$Z = wL_1 = [(c_r + 1)/c_r Z_o]^{0.5}$$
$$w = [(c_r + 1)/c_r w_o]^{0.5} \quad \text{(rad/s)}$$
$$w_o = 1/(L_r C_a)^{0.5} \quad \text{(rad/s)}$$
$$Z_o = (L_r/C_a)^{0.5}$$
$$c_r = C_s/C_a$$

It is important to remember that if $c_r > 1$ then the final voltage of C_s at $wt = 3.142\ (\pi)$ will be positive. But it requires that C_s should be completely discharged.

Therefore, c_r should be less than or equal to unity. With c_r less than or equal to 1, the final voltage across C_s will be negative (or will try to be). The value of this negative voltage will be nominally less than 1 V and will be prevented from exceeding this nominal value by the clamping action of the snubber diode.

An example of a non-dissipative snubber is given in the schematic of Figure 3.3. A description of its operation is as follows.

52 OVER-VOLTAGE PROTECTION

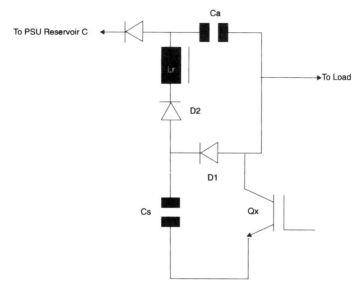

Figure 3.3 Schematic of non-dissipative snubber

Assume Q_x to be fully on at first. Q_x turns off and the load current is diverted into C_s. When Q_x is again turned on the plate of C_a — which was connected to the drain of Q_x — is now connected to the negative plate of C_s. C_s commences to charge, by resonant action, C_a through L_r. By the time it takes for Q_x to be fully turned-off (after being on), all the energy in C_s should have been dumped into C_a. This implies that C_a and L_r should be scaled such that the minimum 'on-time' of Q_x is catered for. When Q_x once again goes through its complete switch cycle the snubbing action of C_s is again invoked, but at the same time C_a discharges into V_{supply} through D_a.

3.4 SOFT CAPACITOR CLAMPS

The purist may well insist that capacitive 'turn-off snubbers' should be grouped within this category; and in the strictest sense there is probably no real distinction. However, I propose to introduce a distinction on the grounds that the voltages arising across the capacitive component (of snubbers) are totally circuit oriented and may well result from the need to reduce dissipation, whereas the capacitive clamp is used solely for the purposes of limiting the voltage across the MGT to less than the break-down value.

One further aspect of distinguishing between the two 'separated categories' is that snubbers tend to be discharged by their associated power switches, whereas capacitor clamps tend to function by *leaking away* the charge on the capacitor.

3.4 SOFT CAPACITOR CLAMPS

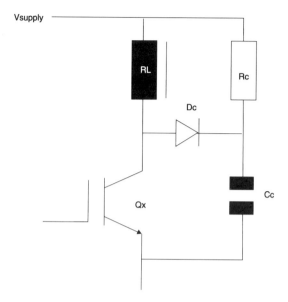

Figure 3.4 Schematic of soft capacitive clamp

Various names are given to the soft capacitor clamps, which are frequently used. Some of these names are:

(1) Spongy clamps
(2) Soft clippers and others.

A close scrutiny of the two circuits, namely snubbers and clamps as distinguished by this part of this book, will reveal some subtle differences.

Examination of the drain–voltage waveform will show that the voltage rise time t_{rv} of the MOSFET is contoured by the snubber, but is not affected to any significant degree by the clamp, until the voltage level of clamping has been achieved.

One type of 'soft' capacitive clamp is depicted in Figure 3.4. In the circuit, as shown, the 'pull-up' resistor R_c ensures that C_c is initially charged to V_{supply}.

When Q_x 'turns on', diode D_c will be reverse-biased. Upon Q_x turning off the voltage at the drain will rise at an initial rate which is determined by the available gate current and the Miller capacitance (C_{rss}) of Q_x, the drain current and the output capacitance (C_{OSS}) of Q_x. This initial dv/dt is maintained until the drain potential of Q_x exceeds that of C_c, and D_c becomes forward biased.

If C_c is made large enough, it will be found that the drain voltage is effectively clamped to some finite value, the delta V_{ds} being determined by the remaining (average) value of I_d and the time for which it continues to flow and also upon the value of C_c.

OVER-VOLTAGE PROTECTION

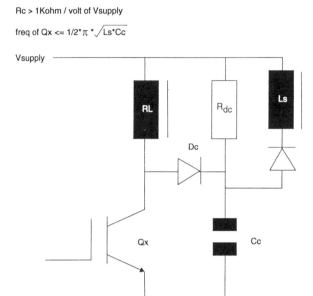

Figure 3.5 Schematic of non-dissipative soft clamp

On the other hand, if the value of C_c is not made large enough to effectively 'clamp' the drain of Q_x, the final value of V_{clamp} (V_{fc}) could be higher than BV_{dss}. The final clamp voltage V_{fc} can be found from:

$$\hat{\imath} = C_c \delta V_{\text{ds}} / t_{\text{off}}$$

where δV_{ds} is equal to $V_{\text{fc}} - V_{\text{supply}}$, and t_{off} is the t_{fi} of the MGT.

As with snubbers, soft capacitor clamps may be either dissipative or non-dissipative. The circuit of Figure 3.4 is the most simple form of dissipative soft clamp. In the circuit of Figure 3.4 the discharge/charge resistor R_c can be calculated from:

$$R_c = t_{\text{off(min)}} / (5 C_c) \tag{3.11}$$

The value of R_c should be calculated such that C_c either fully charges or discharges in the time available. The power dissipated by R_c can be calculated by using the same formula as used for calculating the power dissipated within the snubber resistor.

A non-dissipative soft clamp circuit is shown in Figure 3.5. It should be noted that the discharge of C_c is similar to the discharge action of the snubber capacitor in the non-dissipative snubber of Figure 3.4. Because of this similarity no description of the circuit is really necessary. Since the resistor R_{dc} is only used to charge C_c the power dissipated in R_{dc} will be half of the value obtained by

3.5 CAPACITIVE CLAMPS AND SYMMETRIC PUSH–PULL CONVERTERS

the procedure outlined in (3.11). It should also be remembered that c_r in (3.10), the ratio of C_c and C_a, should be greater than unity.

3.5 CAPACITIVE CLAMPS AND SYMMETRIC PUSH–PULL CONVERTERS

The symmetric push–pull converter (so-called because of the double-ended transformer connection) requires being separately covered in terms of collector/drain clamping against voltage excursions of an excessive nature. Since there are fundamentally two variants of this type of converter topology (current-fed and voltage-fed) it becomes necessary to consider the clamping of the switches for both variants. The two variants will be found to require widely differing clamps, especially if the clamps are of the non-dissipative variety, since the clamp voltage and the potential of the *circuit node* to which the energy is returned could well be incompatible. The reason for this disparity is that current-fed push–pull converters tend to have a pre-regulator stage; with the consequential reflection of the transformer centre-tap voltage being much less than a factor of two lower than the original voltage source in value.

Figure 3.6 illustrates a dissipative clamp for use with the symmetric push–pull converter. The reader should be aware that the capacitor only has the on time of either Q1 or Q2 during which to discharge through R1. The value of C_s can be found by using equation (3.7).

The power loss in R1 can be found by the discreet use of equation (3.7) given below. It should be noted that v_{CLAMP} in equation (3.11a) should be less than

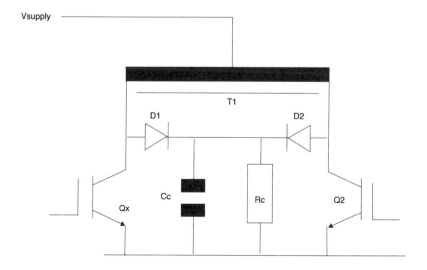

Figure 3.6 Dissipative voltage clamp for symmetric push–pull converters

56   OVER-VOLTAGE PROTECTION

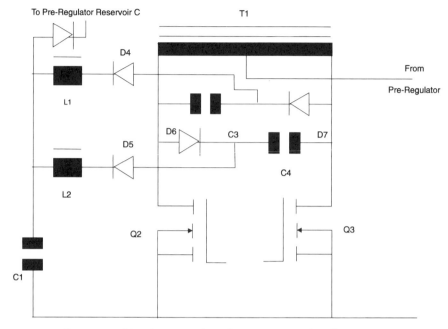

Figure 3.7 Non-dissipative clamp for symmetric push–pull converters

BV_{DSS} for the MGT; and replaces BV_{DSS} in the use of equation (3.7) as it is applied here:

$$\text{Power of R1} = v_{CLAMP}^2/(9 \times R1) \qquad (3.11a)$$

The circuit of Figure 3.6 illustrates the type of clamp which has found significant popularity with symmetric voltage-fed variants of push–pull converters. This clamp may be either dissipative or non-dissipative depending entirely upon the choice of the designer.

The clamp circuit of Figure 3.7 is required if a purely non-dissipative variant of clamp for current-fed converter variants should be desired. This version of clamp is easily configured with the current-fed circuit variant owing to the requirement for conduction overlap.

The discharge of the two clamp capacitors into the original power supply voltage source can be accomplished in either one of two modes — resonant or non-resonant — and entirely depends upon the personal preference of the user.

3.6 TRANSIENT VOLTAGE SUPPRESSORS

There are two types of product that fall within this category and both types may be considered for use as over-voltage protection transducers in circuits where the MGT is required to have some form of drain–source protection. The actual selection of the type of surge suppressor to be used will, to a large extent, depend on the MGT it is protecting. The materials used in the manufacture of these two products are as diverse as the way in which they function.

One other factor which will determine the desirability of the two types, as to which type will finally be used, is the relative cost of the two products.

3.7 VOLTAGE DEPENDENT RESISTORS

Sometimes referred to as Metal Oxide Varistors (MOV), Voltage Dependent Resistors (VDRs) may well be used to protect MGTs with or without an in-built capability to protect themselves to a greater or lesser extent, by way of avalanche. It will be obvious that the MOV should be considered as being essential when the E_{as} capability of the MOSFET is insufficient to cater for the energy in the transient. In this application the avalanche capability of the MOSFET is only the secondary form of over-voltage protection. The MOV is the main component of protection.

The characteristic of the MOV is that they exhibit a high value of resistance until their threshold voltage level is reached. The value of resistance now reduces to a relatively low value, but not to zero, diverting the current in the transient through itself. From data it will now be seen that the finite slope resistance of the MOV (VDR) can be demonstrated, since the clamp voltage, at the level of current flowing in the VDR, will be found to be significantly different to the threshold value. It is extremely important that this behavioural tendency is borne in mind, since the low cost of the component tends to favour its use over the other type of suppressor whose slope impedance is usually considerably lower than that of the MOV.

The relatively large slope impedance of the MOV is the cause of using the MOSFET's avalanche capability as secondary protection. The limited avalanche capability of IGTs may well preclude the use of the MOV for over-voltage protection.

On the other hand, if avalanche capability of the MOSFET is not usable or the switch is an IGT and the choice of transient suppressor remains an MOV then the slope impedance should become part of the design. This accommodation of slope impedance may well mean an increase in the size of the MOV and thereby reduce the considerable price advantage the MOV enjoys.

3.8 TRANSIENT VOLTAGE SUPPRESSOR DIODES

Also referred to as *transzorbs* these devices are in reality avalanche diodes with the ability to absorb very high levels of energy, in relation to their average power rating and size. The difference between this family of parts and the MOV is that these semiconductors exhibit a much lower dynamic resistance by way of comparison with the slope resistance of the MOV, as previously described. It is this 'hardness of clamping' that confers upon the product a greater degree of desirability, over the MOV, for the protection of MGTs with little or no avalanche capability of their own. It goes without saying that its higher cost will be a factor that must be accounted for in the final component selection.

Qualification of final cost means that the cost equation must include the higher cost of the 'rugged MOSFET', which in turn should be offset by the higher price of the Tranzorb.

3.9 GATE CONTROL

Although not being directly concerned with over-voltage protection, it should be remembered that control of the turn-off speed of the MOSFET can ultimately lead to control of the drain–source voltage, and its magnitude because of the $l_{stray}\,di/dt$ effects.

It is for this reason that this section is included in this chapter. Within this grouping are two further sub-groupings: (a) Active clamping, (b) Turn-off speed control.

It is important to repeat once more that too rapid a turn-off excursion of the MGT, especially with an inductive load, can result in extremely high di/dt values. These extremely high di/dt levels will generate excessive levels of transient drain–source/collector–emitter voltages.

It is acknowledged that active clamping has been referred to earlier in this chapter. It is included here as part of the gate control section which is one way of controlling the drain voltage and also the gate voltage, as has been previously referred to.

The use of active clamp circuitry is possibly the optimum approach for large junction discrete MGTs, and are certainly advocated as one of the options for MGT modules. Unfortunately active clamping may not be justifiable for small discrete devices or for low levels of I_d/I_c and power dissipation.

The option which may prove to be the most attractive is 'turn-off speed control'. This is quite easily accommodated as displayed in the circuit in Figure 3.8.

The resistor in series with the diode should be selected for optimised 'turn-off' V_{ds} control.

Gate control of di_d/dt is especially important when MOSFET modules are 'turned-off'. The reasons associated with the previous sentence arise because of the *internal* inductances of the main power *terminals*.

It is therefore very important that any data-sheet information or limits relating to di_d/dt as a function of supply voltage be carefully studied. Consider the ramifications of the following scenario involving the use of a module housing several parallel-connected MOSFET junctions.

3.9 GATE CONTROL

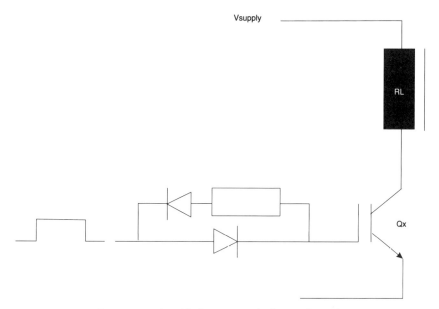

Figure 3.8 Simplified gate control of turn-off speed

MOSFET turn-off of the module will result in some finite drain–source voltage which for these purposes may be termed $V_{DS(EX)}$, prior to current fall-time initiation. The voltage created by the di_d/dt by the internal inductance of the module should be summed to $V_{DS(EX)}$. If the sum of these voltages exceeds the BV_{dss} of one of the devices then avalanche break-down of that particular device could well be the result. By itself this would not necessarily be dangerous — except that the single junction may well have to sustain the full drain current of itself and also the currents of all the 'bed fellows' under the stress of avalanche. Ultimate damage to that particular junction could well be the result. In the case in question it is patently obvious that prudent turn-off speed by gate control should yield the optimum conditions.

More sophisticated circuits will definitely confer a very worthwhile improvement in the accuracy of control. The reader should temper the urge to obtain greater accuracy of control, with the realism of whether the improved accuracy can be justified.

If the superior accuracy is justified then the gate-control circuit of Figure 3.9 can be just one of the many which may be examined.

The circuit utilises the principle that voltage rise time t_{rv} and current fall time t_{fi} are not independent of each other, but are in effect interdependent as determined by the gate-charge control characteristics of the power switch being used. The rate of gate-charge extraction for Q_x will be set by the 'current sink' impedance

60 OVER-VOLTAGE PROTECTION

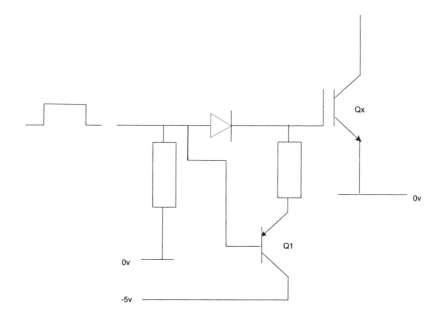

Figure 3.9 Schematic of drive circuit with charge control

of Q_1, and this equates to being the ratio of R_{DC} and the h_{FEk} of Q_1. The negative gate bias (set by V_{OFF}) is essential in order that Miller turn-on of Q_x is not possible. (The ubiquitous qualities of the circuit of Figure 3.9 will be demonstrated in later chapters.)

A secondary benefit provided by the use of turn-off control, will be a significant reduction in the generation of EMI/RFI.

3.10 CIRCUIT LAYOUT

The principles of good circuit layout are:

(1) Maintain all conductor lengths as short as is physically possible in order to reduce the levels of stray inductance.
(2) Currents in all conductors should be maintained as continuous, if possible, with a small riple component.
(3) Gate circuit return leads should connect directly tot he source/emitter terminal of the MGT package in order to eliminate the effects of feed-back from L_{SOURCE}.

This theme will be returned to frequently, and no excuses will be offered by me for this repetition. It cannot be stressed too much that this is possibly the most vital of practical procedures which should be incorporated by the reader especially if the primary power switch should be a power MOSFET. This most important of all circuit constituents will overcome secondary problems which may arise if this precaution was to be overlooked. Some of the benefits which may accrue from good circuit layout are:

(1) Elimination of the interference at source. This aspect is much more preferable than the filtering out of the interference, since it is certainly more cost-effective.
(2) Improved levels of performance and efficiency.
(3) Improved reliability arising out of the reduction of stresses which may be applied to other devices or components.
(4) Optimised performance due to good layout can easily lead to the use of relaxed filter requirements with the possible reduction in overall cost. This last statement may appear to be little more than repetition of (1) but in reality is much more profound. Even if it was merely a repetition of (1) the importance cannot be overstated.

3.11 TOPOLOGICAL CLAMPING

Although not entirely related to this chapter, the insertion of this short section arises out of the need to explain that topology can play a significant part in protecting the MOSFET from over-voltage. This is particularly true for power supplies where certain converter topologies are more effective than others at providing a measure of automatic clamping, and will be more fully dealt with in a later chapter.

It would be presumptuous to recommend any of the voltage protection methods outlined in this chapter. The one which the reader should adopt may well be decided upon after consideration of several factors. It should not be overlooked that one of these factors could be that of personal preference, or one that the reader feels most comfortable with. There can be no hard and fast rules which can be applied in order that selection of protection methods can be made any the easier. The purpose of this chapter is merely to outline the options available and to then leave it for the reader to decide.

4
Over-current Protection

Although the power MOSFET and more recently the IGBT, have deservedly earned their reputation for ruggedness and have the ability to handle considerable overloads, for moderately short durations, where current is concerned it is still considered prudent to ensure that the MGT is protected adequately from excessive current stress. This protection may well be in addition to any form of current limiting that may be used for circuit protection under normal operating conditions.

It is strongly recommended that high rupturing capacity (HRC) fuses are not used to protect MGTs from current surges.

HRC fuses are well known for having fusing currents which may frequently be three or four times greater than their rated current. Moreover, the clearing times for these fuses, after element fusing, are ill-defined to say the least and as such infer that MGTs would probably have failed long before the fuse will have cleared.

It is also advisable to ensure that the IGT under consideration is completely latch-proof. Some devices from certain vendors are completely latch-proof, but this 'freedom from latching' does not always apply. Therefore, *test and be certain* should be a golden rule.

It is strongly recommended that circuit layout and component design procedures observe strict attention to detail, since the design of these components can affect the currents which flow through the power switch. The dividends reaped, because of these recommendations being stringently observed, will be handsome.

For example, consider a switch mode power supply where the transformer primary winding (due either to neglectful design or poor quality control during manufacture — or an unfortunate combination of both) is found to have an extremely high 'shunt (or winding) capacitance'. The consequences are that during every switch cycle the MGT switch will have to handle a spike of current (of such a magnitude as to possibly over-stress the MGT or some other component). Obviously, current protection techniques can be used, but one should first consider all of the other possible effects. Circuit complexity has been increased and quite possibly unnecessarily. System cost has to some degree been adversely influenced. It would be quite possible to discover that some aspect of stress is still in existence and as a result reliability is found to remain unchanged. A further variable which could cause current spikes is inadequate allowance being made for remanence in the design of the transformer — if the SMPS design uses a forward or fly-back topology. Push–pull converter transformers should have allowances made for *flux doubling* at start up. How much better it would have been for the diligent designer if the original design excellence had extended to the transformer and winding design.

If all the precautions relating to transformer winding design have been found to have been observed and implemented, and an over-current problem is found to persist, then other causes of the over-current problem should be suspected and other forms of protection must now be investigated.

The techniques which will be covered in this chapter will cover such areas as:

(1) The use of special MOSFET device structures.
(2) The use of special circuit techniques for current limiting or tripping-off after detection of an over-current incident.
(3) The protection of the parasitic transistor, after being used as an inverse-parallel diode (from high values of dv/dt), during diode recovery.
(4) The causes of simultaneous conduction in half and full-bridge converters and also within voltage-fed symmetric push–pull converters.

4.1 CURRENT MIRROR/SENSING MOSFETS

This is the generic name that is applied to all types of power MOSFET that utilise some form of current mirror which is created within the junction during fabrication. This current mirror is constructed through choice and not by some quirk of fate, or through chance. Various brand names have been given to these unusual devices and include *HEXSENSE, MIRRORFET* and *SENSEFET*. The three names are applied to virtually identical structures and therefore may well be considered as being interchangeable within certain limits. It is advisable in the case of the current sensing MOSFET that selection criteria include the sense ratio r besides the usual ones of on resistance and current handling capability. Care should be exercised in not confusing the device known as the *TEMPFET* within this overall classification as being a current-limiting MOSFET with some form of thermal trip being incorporated for added protection.

The TEMPFET is in reality a standard MOSFET with a *temperature sensitive thyristor* bonded to the die of the MOSFET. The designed objective for this product is to sense the junction temperature, which may well result from excessive current flow, and to then *crowbar* (apply a switched short-circuit) the gate-drive and so turn off the device. It may, therefore, be regarded as being somewhat analogous to a current-protected MOSFET, but with one serious built-in limitation. The device is not protected from currents in excess of its published

4.1 CURRENT MIRROR/SENSING MOSFETS

limit for pulse operation I_{DM} (when short-circuit of the load may occur — this being a mandatory requirement for some types of motor control circuits) and this may prove to be a serious limitation for this particular type of device.

As indicated in the previous chapter other forms of *SmartPower* (a generic name applicable to certain types of monolithic integrated circuit) device with in-built current limiting and over-temperature protection are available. However, since many of these devices may well include a power BJT as the output power device, their inclusion will be neglected since the design-in of BJTs, as a general rule is not one of the objectives of this book. Another objective to the inclusion of design-in information of SmartPower devices is that by their very existence they obviate to some extent the requirement for design expertise. The introduction and use of these devices will proliferate in the way that logic ICs have done, thus making it all the more desirable. I would strongly advise the reader not to wait too long for these devices to over-run the market-place. The use of these devices will initially extend to certain niche applications and will gradually widen their area of use. Standard devices will continue to be the mainstay for some considerable time.

The device whose schematic symbol is outlined in Figure 4.1 is a true current-sensing MOSFET in every way and may therefore be used within the circuit for the additional purpose of being able to measure current. It is created

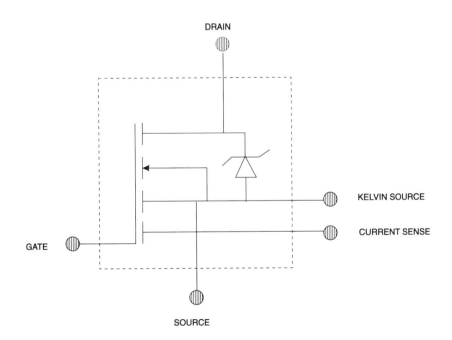

Figure 4.1 Circuit symbol of current-sensing MOSFET

by simply isolating a few of the cell-sources from the many thousands of parallel connected individual cell-sources that constitute the make-up of the junction.

The cell-sources thus isolated during wafter fabrication have their own separate terminals for the sense and non-power source terminals. This last terminal is often referred to as a Kelvin terminal. This tiny device (the current mirror within the power junction), a perfect MOSFET in its own right, mimics the action of the remaining cells of the junction — but in a unique fashion.

The drain current of this small MOSFET is said to mirror the current flowing into the drain of the MOSFET in its entirety, but in direct proportion to the ratio of the cells which go into the make-up of the overall device. (Note that true mirroring only occurs during operation of the MOSFET in the linear region, and in a slightly modified manner during operation in the saturated region.)

The sense ratio r can be equated thus:

$$r = \frac{\text{Total no of cells}}{\text{No of isolated cells}} \qquad (4.1)$$

The term r is used to define the sense ratio and is in turn used to indicate the current which may be expected at the sense terminal, for a given current which flows into the drain terminal.

During operation in the saturated region the sense ratio r is found to equate to the ratio of on resistances. The ratio may therefore be expressed as:

$$r = \frac{\text{Total } R_{ds[on]}}{\text{Sense } R_{ds[on]}} \qquad (4.2)$$

The output of the sense terminal is capable of being utilised for several interesting applications.

It should be understood that the current-sensing MOSFET is incapable of performing current-limiting functions by itself. Its sensing capability must be used as part of an overall circuit function.

The areas to which the best use of the current-sensing MOSFET can be put are obviously power supplies (both switcher and linear); automotive (ignition, fuel injection and high-side switches) and numerous others. The limit is confined solely within the imagination and ingenuity of the reader.

The most obvious way of demonstrating the use to which the current-sense MOSFET can be optimally used is in the basic linear voltage regulator, which must possibly rank as the most ubiquitous of power supply circuits.

One of the reasons for the immense popularity of this circuit is its relative simplicity. Because of this simplicity of design no attempt will be made to pursue the subject of the regulator itself. The author will instead concentrate purely on

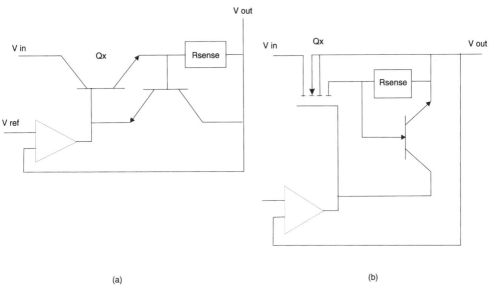

Figure 4.2 Simplified schematics of linear current-limiting using BJT (a) and current-sensing MOSFET (b)

the option of current limiting. The easiest way to visualise the benefits which may accrue out of the use of the current-sense MOSFET is to directly pit it against the BJT in the linear voltage regulator. The way the two circuits of Figure 4.2 operate are fundamentally the same. Increasing load-current through Qx causes an increasing potential difference to be developed across resistor R_{SENSE}. When the base-emitter voltage of the NPN transistor Q1 is sufficient to turn Q1 on, both the base (and the gate) of the series-pass device Qx will experience drive starvation and current limiting is said to have commenced. Any attempt at further increasing the load current will only result in further drive limitation. It is now apparent that the regulator has transited from constant voltage operation to that of constant current.

The significant difference in achieving this transition is by the use of a power resistor for R_{SENSE} in the case of the BJT, whereas the resistor for the MOSFET can have a dissipation capability of less than one-eighth of a watt if the design is correctly made. Fold-back or re-entrant current limiting can be accommodated with equal alacrity.

In both instances the common variable for determining current limit is V_{be}.
In the case of the BJT:

$$I_{C(\text{limit})} = \frac{V_{be(Q1)}}{R_{SENSE}} \quad (4.3)$$

Whereas for the current-sense MOSFET:

$$I_{D(\text{limit})} = \frac{rV_{be(Q1)}}{R_{\text{SENSE}}} \qquad (4.4)$$

Fortunately the use of the current-sensing MOSFET is not restricted for the solitary application of current limiting in linear regulators. It can be used with equal aplomb in switchers. The inclusion within switchers can be extended to three distinct modes of current measurement and control. These modes are:

(1) *Linear current limit/control.* In this mode, the operation of the current-sense MOSFET may be considered as being similar to that of the linear regulator, i.e. when a current threshold is exceeded the switch element is brought out of saturated operation, by gate drive starvation, and put into its linear operating region. The major disadvantage with this approach is the very high level of dissipation sustained by the MOSFET during operation in its linear region.
(2) *A pulse by pulse current-time mode within PWM operation.* The sense terminal is now connected to the current-limit input of the pulse-width modulator comparator. When the current limit threshold is exceeded the PWM comparator will terminate the gate-drive pulse to the MOSFET.
(3) *Pulse current-trip outside of the control-loop.* This mode of over-current protection is similar to that of (2) above, with the exception that the gate-drive signal is terminated from outside of the control loop. Care should be observed in the utilisation of this approach in order that loop stability be maintained.

All of the techniques are particularly suited to high-side switch configuration for the automotive industry. This enables the so-called smart high-side switch to be fabricated with $R_{DS[on]}$ values which may be considered as being economically impractical by monolithic solutions.

A circuit which will perform the current-trip function is displayed in Figure 4.3. The silicon controlled rectifier Q1 which is connected between the gate and the Kelvin terminals of Qx should be of the gate sensitive type. This device has a very low average current rating, since the only role it fulfils is to discharge the input capacitance of Qx. The operation of this circuit is self-explanatory and will therefore not be discussed.

The reader is reminded that the gate-sensitive SCR must have a very low value of holding current. This requirement results from the fact that once the input capacitance is discharged, the available current, which would normally be available for once again charging the input capacitance, should be diverted through the SCR, which should remain latched.

The examples which have been demonstrated so far cannot be considered as having exhausted the circuits which the user should be able to develop with the adoption of a certain degree of ingenuity.

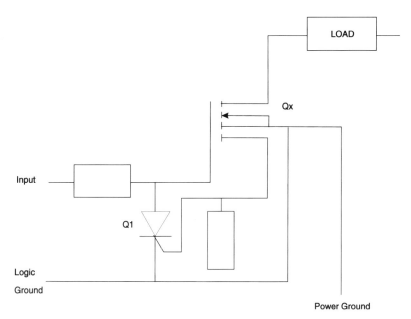

Figure 4.3 Current-trip circuit with mirror MOSFET and SCR

4.2 CIRCUIT TECHNIQUES FOR OVER-CURRENT PROTECTION

The current-sensing MOSFET is not the only device which should be considered, although it would be reasonable to say that it is certainly the easiest of products to protect from the stress of current overload.

The conventional MOSFET or IGT with its extraordinary pulse-current capability may well be the only part that may be useful in a particular circuit. The I_{DM} rating should not, however, be misconstrued as bestowing immunity against failure from current overload. Over-current protection for the device should be considered as being one of the facets of good design.

Protection of the MOSFET as part of the overall circuit design, apart from the use of current-sensing devices, can be made from one of many alternative choices. These choices can range from circuit control of drain current as part of the overall scheme of things, or it can be localised to merely protecting the power switch, by itself. Whichever option is decided upon, the measurement of current must have over-riding priority.

4.3 CURRENT MEASUREMENT

Before it is possible for current protection to be invoked, it is essential that the current itself should be measured with a certain degree of accuracy. The methods by which current measurement may be achieved fall into the three main categories below:

(1) Linear detection.
(2) Resistive measurement.
(3) Inductive measurement (by transformer).

Linear detection

The term 'linear detection' is applied to the method of measuring the voltage, developed across the MOSFET itself, as being a measure of the drain current. The theory behind this approach is perfectly sound, since the standard MOSFET demonstrates the qualities of a resistor in the saturated region of operation.

The reader is advised to be aware the effect junction temperature has upon the $R_{DS[on]}$ of the MOSFET. Increases of more than 100% are possible for t_j excursions from 25 °C to $t_{j(max)}$. The effect of temperature upon $V_{cs[sat]}/V_{ce[on]}$ for the IGT is far less pronounced and may under certain conditions (when operated in the linear region) fall with a slight increase in current. In both instances it is advisable that the $V_{ds[on]}/V_{ce[on]}$ threshold be made for a compromise value, thereby offering reasonable protection over the full temperature range. It is possible that some compensation can be made as will be demonstrated below.

Those readers who have experienced the use of linear detection with BJTs will find no surprises when applying this method of current measurement to IGTs.

Having achieved the actual measurement of drain current we must now decide how best to manipulate this information.

The first contingency which must be allowed for is the isolation of the remainder of the control circuit from the collector/drain of the power device itself; since the potential at this terminal may well prove to be catastrophic to the rest of the control circuitry, this is especially true if the power circuit is part of an off-line application.

The diode D1 in the schematic of Figure 4.4 must of necessity be of the ultra-fast type, whilst at the same time being able to block reverse voltages of up to 400 V or more.

The forward current capability of D1 can be negligible. The way the circuit operates is as follows: when Qx is switched on the voltage across it (v_{Qx}) will be determined by the current flowing through it. This voltage when summed with the forward voltage of D1 is compared with the reference voltage v_{ref} by the comparator IC1. If the sum of these two forward voltages, namely that D1 and V_{OX}, exceeds V_{ref} the output of IC1 will change its state and ultimately lead to the termination of the gate-drive of Qx. The reader may well query the desirability of linear detection on the grounds of the variables introduced into the measurement and control loop. These variables being the change in $R_{ds[on]}/V_{cs[sat]}$ of Qx with the respect to temperature; and to some extent compensated for by the change in V_f of D1 with respect to temperature.

4.3 CURRENT MEASUREMENT

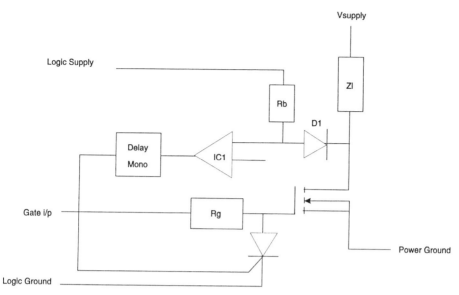

Figure 4.4 Circuit of linear detected MGT and control circuitry

It is arguable, with some justification, that the V_f change will compensate for the change in V_{Qx}, but the compensation may prove to be sufficient. All of these premises are valid and it is a sad reflection, on the part of some IC designers, that the technique has been adopted with sufficient alacrity to warrant PWM ICs including an input for anti-saturation detection.

Series resistor or current shunt

The adoption of this method of current measurement may be considered to be a retrograde step in view of the availability of the current-sensing MOSFET. The justification of this attitude can be on the grounds of the adverse power dissipation in the series resistor. The reader should remember that this was one of the more cost-effective solutions before the advent of the current-sensing MOSFET.

The advantages of adopting this type of measurement transducer are considerable and can be listed as:

(1) Cost (relatively low if moderately toleranced parts are acceptable).
(2) Simplicity.
(3) Availability (there is an overwhelming preponderance of suppliers for such a component.)

The disadvantages associated with the use of the current shunt are in many ways as numerous as the benefits. The moderately low cost of the component (if tolerances of 5% and 10% are accepted) rapidly becomes less obvious if lower toleranced parts are specified.

Many engineers forget about the inductance of these components. When they are reminded they will counter this criticism by specifying non-inductive parts and further negate any price advantage which existed. Consider the chagrin with which the alleged 'non-inductive' part is greeted when it is found to have less inductance than a standard resistor. This inductance, when coupled with any stray capacitance in the immediate vicinity of the resistor, results in an oscillatory circuit whose natural resonant frequency may prove to be the perfect match for the MOSFETs inherent speed. The resultant ringing will degrade any accuracy which may have been hoped for and the 'slugging' or filtering which must now be inserted may result in reduced response at best. At worst is the distinct possibility that the current loop will be conditionally stable if the designer is only moderately unlucky.

The power dissipated within the shunt, no matter how insignificant it may be to the user, can invariably become one of the major causes of unreliability, since heat is one of the major influences of failure.

In spite of these undesirable features it is to the credit of engineers that this technique has proved to be so deservedly popular.

This option should be considered with the proviso that all the disadvantages are allowed for; the ensuing circuit could well be found to be cost-effective with the added bonus of being quite easy to configure.

Current transformer

The use of the current transformer has proved to be extremely popular in circuits where high voltages are often encountered, or where isolation barriers have to be crossed. Latterly, 'Hall-effect devices' are being considered as a potential replacement for this component, but until this becomes a *de facto* reality the reader should be aware of the potential pitfalls.

With the vast majority of transformers leakage inductance should be minimised at all costs. This philosophy should be applied to the design and construction of current transformers with equal vigour. The warnings relating to the inductance of shunts will prevent similar consequences occurring when current transformers are included in the power circuit.

Invariably the current transformer core will be a toroid. The cautionary advice given here can be applied to all other core shapes. The generation of leakage fluxes is the prime source in the creation of leakage inductance and the generation of these fluxes must be strictly avoided. Certain core shapes do not lend themselves to leakage flux containment as easily as one would wish. Fortunately the toroid is not one of these cores.

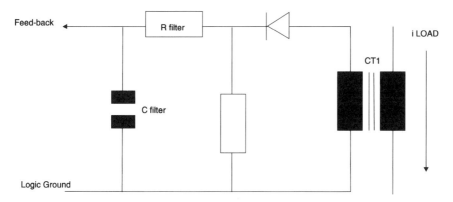

Figure 4.5 Current XFMR with secondary rectification

The minimisation of leakage inductance should be conducted as follows:

(1) The core should be covered in its entirety wherever practicable. It may be completely impractical to do so with the primary winding — especially if this winding is confined to being a single turn, or to being comprised of no more than a few turns. The secondary, on the other hand, may well not have this restriction imposed upon it.
(2) If total coverage of the core is not possible then the winding with the greatest number of turns should have its individual turns spaced equally over the surface of the core, while surrounding the other windings where possible and thus isolating these windings from any leakage fluxes.

Having established some guides to the construction of current transformers, it is opportune to consider the overall circuit into which the transformer will be installed. If the secondary signal is an alternating current and it is desired that this signal should be rectified, then it is vital that the rectifiers be included within the circuit before signal conditioning is undertaken. The effects of temperature upon the forward volt-drop (V_f) of the rectifier diodes will have been allowed for because of their inclusion within the overall measurement circuit. This type of circuit is shown in simplified form in Figure 4.5.

4.4 TURN-ON SNUBBERS

So far this chapter has covered the measurement techniques used for the supervision and protection of the MOSFET from current overload, where these overloads are mainly related to the conventional elements which are easily identified within circuits. Virtually no thoughts have been given to those elements which may well be hidden — or forgotten.

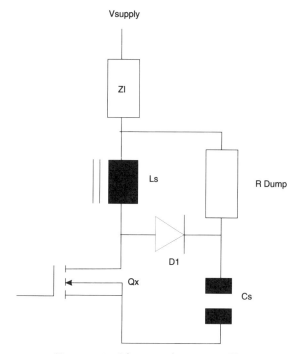

Figure 4.6 Schematic of turn-on snubber

These elements are usually capacitive in nature and have the tendency of stressing the MOSFET quite severely during the turn-on phase, especially if switching speed is of prime importance.

These capacitive components are usually the 'shunt' capacitance of transformer windings or the windings of inductors. The 'inrush' current associated with this type of hidden element can have unforeseen consequences. It is therefore advisable to cater for such elements.

One other possible reason for the inclusion of a turn-on snubber may well be because of coincident conduction in push–pull types of circuit. The prevention of coincident conduction during the design phase should be of primary concern with due allowance being made for dead-time. Unfortunately gate control, or the lack of it, is not the only means of inducing coincident conduction and will be considered in a later part of the chapter.

One type of turn-on snubber is an inductor connected in series with the load, it is not important if the load is a transformer or not. The purpose of this secondary inductor is to limit the di/dt of the 'switch current' to a safe value during the charging of the capacitance of the main inductive component. Such a circuit arrangement is shown in the circuit in Figure 4.6. It should be noted that the turn-off snubber should encompass the turn-on snubber as well.

4.4 TURN-ON SNUBBERS

The winding of the turn-on snubber coil should be fabricated in such a manner that its own 'self-capacitance' is minimised. One method would be to use a pile winding approach.

The first choice is to obviate the use of the turn-on snubber by ensuring that the transformer or inductor winding is of minimal capacitance in the first place. Unfortunately this is not always possible — especially if the transformer or inductor happens to be a bought-in component.

One other component which falls within this context is the barrier capacitance of the Schottky rectifier. Since this component is usually associated with power supplies its effects will be studied in greater detail in Chapter 8.

The calculation of the value of the turn-on snubber is not difficult — nor should it be. The frequent solution, however, is to arrive at its value by a process of trial and error.

In order that the reader is not tempted to follow the above mentioned trial-and-error procedure; the determination of the turn-on snubber is outlined below:

$$e_{\text{eff}} = L_s \, di_l / dt_{\text{sw}} \qquad (4.5)$$

where e_{eff} is the voltage that is applied across the winding being 'snubbed', L_s is the value of the snubber inductance, i_l is the load current and equates to the final value of the drain current which results from the load, and t_{sw} is the desired value of current rise-time.

The topics which have been discussed so far have all been concerned with protecting the MGT from the possibilities of overload. In this case it is the channel conduction which has been considered. The MOSFET structure includes a parasitic BJT which constitutes the inverse parallel body-drain diode. No consideration has been accorded to this part so far in this chapter. Nor has any consideration been given to the parasitic SCR in the IGT.

The major problems concerned with over-current protection associated with the parasitic diode is that of i_{RM} (the peak value of reverse recovery current). Exceptionally high values of i_{RM} are usually due to dv/dt turn-on of the parasitic. This parasitic turn-on was a frequent source of failure in early power MOSFETs. Fortunately this problem has been considerably alleviated in present devices. Since the parasitic diode is an important element in inverters associated with motor control — especially where PWM is utilised — the failure mechanism associated with the parasitic, and alleviation of the stresses due to i_{RM} will be discussed in greater detail in Chapters 7 and 9.

Coincident conduction is not confined to dv/dt induced turn-on of the parasitic. Nevertheless dv/dt induced turn-on can still be a major source of failure or poor efficiency. The mechanism is turn-on of the channel itself by the application of high dv/dt across the off MOSFET leading to Miller turn-on. This phenomenon has already been referred to in Chapter 2. Miller turn-on most frequently occurs in totem-pole push–pull stages but may also occur in symmetric push–pull circuits, although the likelihood of this occurrence in the symmetric push–pull can be seen to be diminished.

Coincident conduction problems associated with dv/dt turn-on can be overcome by the judicious use of reverse bias applied to the gate. In all other cases the problem of coincident conduction can be put down to inappropriate attention paid to timing in the small-signal circuit.

A final word of caution on the subject of overcurrent relates to the MGT's *pulsed collector/drain current* rating (I_{CM}/I_{DM}). In the case of MOSFETs it is frequently asked if it is possible to exceed the data sheet value which has a 10 μs limit for a shorter period than specified. The answer is usually that the maximum rating is given for extremely good reasons and not for the vendor's benefit. The I_{CM}/I_{DM} limit ensures that the ohmic voltage drop along the length of the channel does not increase the possibility of emitter injection occurring in the parasitic, leading to parasitic turn-on and its subsequent transit into second-breakdown.

5
Thermal Management

The reader is again advised of the necessity of being aware, when designing for reliability, that one of the variables which occurs within the calculation of Mean Time Between Failure (MTBF) is component temperature. The maintenance of component temperatures beneath specific maximum values is particularly important if MTBF of the equipment is to be optimised. These specific maxima may well be values determined by the manufacturer for active components and sometimes by the law of nature for certain passive components, namely the Curie temperature for magnetic materials.

Exceeding these maximum temperatures is not advised and is done so at one's peril.

Because this book is about designing circuits with MOS-gated transistors, the bulk of this chapter will address thermal management related specifically to these devices.

Rules which have been advocated for subjects covered in previous chapters have been formulated with regard to these rules that have been tried and tested. It is, therefore, advisable that rules should also be formulated with respect to maximum temperatures for the MOSFET junction. There is only one rule of any real consequence. This may be stated as: *never use MGTs in a manner such that the junction may be forced to stray outside of the temperature limits which have been stipulated by the vendor.*

This exhortation is not undertaken lightly. The reasons for working within the confines of the minimum and maximum temperatures are well documented within textbooks and may therefore be considered as being superfluous. The reasoning will, therefore, not be repeated in these pages.

Although the validity of this rule is self-evident, namely, the necessity for staying within temperature limits, it is necessary to justify the demand for this rule and to devise ways of ensuring that these limit values are adhered to as far as is possible.

'Specmanship' has frequently been employed in the production of data sheets and this has never been more in evidence than in the case of data for the BJT. The reader may well be aware of data for some BJTs having maximum operating temperatures as high as 200 °C. At the same time, at temperatures above 125 °C, the crystal lattice of the BJT undergoes subtle changes, resulting in the temperature stability of the junction being seriously jeopardised. The cause of this junction instability is the phenomenon known as current crowding.

The MGT, on the other hand, does not exhibit this phenomenon except during certain periods of operation within the linear region. The implication is that the MGT may well be immune to abuse from thermal mismanagement. Nothing could be further from the truth and it is the purpose of this chapter to ensure that feelings of undeserved ill-will towards the device are not caused by mismanagement of thermal aspects of the user's part.

The primary concern of the first-time user should be an awareness of the MOSFET's inherent tendency to increase its on-resistance as heating of the junction takes place. This positive temperature coefficient of resistance can be distinctly beneficial when parallel connection, of two or more devices, is considered. Unfortunately, this same temperature dependency can be seriously embarrassing if adequate thermal management is not catered for. Therefore always provide adequate heat-sinking.

The second golden rule to consider is: *low thermal resistance heat-sinks can not only improve reliability; they can also improve efficiency and/or reduce cost.*

The question is frequently asked how the thermal resistance to ambient of the heat-sink can be correctly ascertained. This is best done by applying a known level of power and to then measure the temperature rise of the sink. The known level of power can be accurately set by mounting a metal-clad power resistor to the heat-sink and then to pass a precise d.c. current through the resistor. Care is required in the mounting of the resistor to ensure that the thermal resistance of the resistor case to sink is minimised. The use of any one of a number of proprietary thermal greases is recommended.

I am resigned to the vagaries of purchasing officers and their never ending quest for finding the lowest-cost components, without any thought being given to the effects upon reliability which their cost-conscious philosophy induces. It should be pointed out in fairness that I am not denigrating this philosophy, merely pointing it out. The good purchaser performs some admirable achievements in the service of their company. It is none the less important to realise that the purchaser should not be the final arbiter as to which component may be used.

In my experience it is the minority of companies which delegates responsibility to the design engineer to completely specify the components which will be used in the manufactured equipment. I commend these companies.

Unfortunately there is a large number of companies who entrust the purchaser with the overall responsibility for purchasing components which conform to a relatively loose specification.

When this philosophy is applied to active components the purchaser will usually look at the specification, which may in some instances include a part number, and order the part complying with the specification from the vendor quoting the lowest price. If a part number is part of the specification then the same rule is applied with no thought being given to the possibility that the data for the part from the originating supplier may be superior to the part being supplied by the lowest cost vendor. There is also the distinct possibility that the lowest cost component data sheet in no way complies with the data of the part which was originally used for the circuit development. I therefore repeat that purchasers should not be entrusted with the responsibility of deciding which vendor's parts will be purchased in accordance with a specification.

When problems arise due to the inferior part not performing in the manner in which the part used by the designer during the initial design phase did, it is the design engineer who unfortunately tends to be held responsible for the failure. In this instance the well known phrase *caveat emptor* will be found to be particularly applicable.

5.1 $R_{ds[on]}$ VARIATION AS A FUNCTION OF TEMPERATURE

I use $R_{ds[on]}$ variation as a function of temperature to highlight the major aspect of thermal management which should be adopted when initially selecting any particular power MOSFET. First-time users should pay particular attention to the effect of temperature upon this parameter when finalising the circuit design.

The procedure which I advocate is applied from the outset of the design phase:

(1) Make the initial selection from data for the $I_{d(cont)}$ rating given for a case temperature of 100 °C. The data for $I_{d(cont)}/I_{c(cont)}$ for a case temperature of 25 °C is almost worthless unless the equipment is to be operated within a container filled with a Chloro-Fluoro-Carbon (CFC). The 25 °C rating is a 'carry-over' from BJT days, and is still used by some MOSFET/MGT manufacturers for commercial reasons; reasons which can be disregarded as being mainly concerned with 'specmanship'.

(2) Using the data for $R_{ds[on]}$ given for $t_{case} = 25$ °C (always make use of the maximum data sheet value), calculate the power loss due to conduction. Use

$$P_d = i^2_{d(r.m.s.)} R_{ds[on]}$$

Now increase the value for P_d by 25% to empirically allow for switching losses. Let $P_{sw} = 0.25 \times P_d$. This value is accurate enough as a first-cut approximation for switch frequencies f_o up to 75 kHz. For higher frequency switching multiply P_{sw} by the ratio of $f_o > 75/75$. Do not vary this value of P_{sw} for any subsequent reiteration.

(3) Make $P_{tot} = P_d + P_{sw}$, P_{tot} being the total power dissipated by the MGT.

(4) Select a heat sink with a $R_{\theta(sink \text{ to ambient})}$ which will give a t_j rise of 50 °C above ambient.

80 THERMAL MANAGEMENT

(5) With this new value for t_j use the $R_{ds[on]}$ normalisation curve from the data sheet to define the new value for $R_{ds[on]}$.

(6) Now reiterate by repeating steps (2) and (3). With the new figure of P_{tot} and using the R_θ for the heat sink once more ascertain the new value of t_j. If this new figure of t_j is below the data sheet limit for $t_{j(max)}$ go to step (5) and then on to step (6). (If t_j is higher than $t_{j(max)}$ select a new heat sink or select a new device.) It will now be discovered that a maximum of four reiterations should suffice.

If the power transistor is a MOSFET and is to operate in-circuit with its own heat sink then no further consideration is necessary. If the device, on the other hand, must share heat-sinking with other components such as rectifiers, then the effect of heat sink temperature rise upon the dissipation of these components (the MOSFET) must also be taken into account and should also be used in steps (1) to (6) outlined above.

If the power transistor happens to be an IGT then a slight variation to step (5) outlined above will become necessary. In this instance the normalisation curve for $V_{cs[on]}$ should be used.

The full effects of temperature variation for the IGT is given below.

5.2 $V_{ce[sat]}/V_{cs[on]}$ VARIATION AS A FUNCTION OF TEMPERATURE

The initial calculation of step (2) must now be $P_d = V_{ce(sat)} \times I_c$. The steps used to calculate P_{sw} must also be altered. To calculate P_{sw} use:

$$P_{sw} = E_{ts} f_{sw} \qquad (5.1)$$

where E_{ts} is the total switching loss per pulse, and the energy value is to be found in the data sheet for the part, f_{sw} is the switch frequency.

The value for E_{ts} is given for a t_j of 150 °C, and therefore no variation for P_{sw} is required during reiteration.

5.3 OTHER THERMAL MANAGEMENT CONSIDERATIONS

Effects of package outline

Frequently it will be found that a given MGT chip may be available in several package outlines. The need for the various package outlines arises from the need to satisfy a large number of application requirements. Some of these requirements may be as diverse as surface mount requirements necessitating Surface Mount Devices (SMD) or hermetic packaging for military or quasi-military applications.

5.3 OTHER THERMAL MANAGEMENT CONSIDERATIONS

Since the major thrust of this book is devoted to power applications then the packages which will be discussed, will be those relating to significant power handling capabilities.

The fact that the TO-3 (and its sibling the TO-66) has been found to be devoid of any desirable features, except for its hermeticity and the fact that it has been an unquestionable requirement for military use until recently, precludes it from being included here.

The two packages which will be scrutinised most closely will be the TO-247/TO-218 and the TO-220. Both of these plastic packages now have metal variants available for hermetic and military applications.

Let us consider the case of the same die availability in both packages and examine the effects they have upon thermal management and reliability. In this instance the die under scrutiny is the industry standard size 4 and is rated for 500 V blocking capability. The respective part numbers are IRF840 (TO-220) and IRFP440 (TO-247). Both of these parts are rated as follows:

Part no.	$I_{d(cont)}$ @ $t_{case} = 100\,°C$	E_{ar}	P_{tot}	$R_{\theta JC}$	$E_{\theta[CS]}$
IRF840	5.1 A	13 mJ	125 W	1.0 kW	0.5 kW
IRFP440	5.6 A	15 mJ	150 W	0.83 kW	0.24 kW

The variations in ratings for the two parts shown above carry three possible implications:

(1) For all things being equal (as far as heat-sinking is concerned) the TO-247 part can operate at a higher load current.
(2) If the TO-247 part is operated at the same current as the TO-220 part, with the same heat sink, then the t_j for the TO-247 will be lower than for the TO-220 and this will result in (a) higher efficiency (because the $R_{ds[on]}$ will be lower) and (b) superior reliability.
(3) It is possible that a smaller heat sink could be used with the possibility of lower overall cost.

The TO-247 part will inevitably be marginally more expensive than its TO-220 counterpart. This increased price could tempt the designer to pare costs by using the TO-220 device. Regrettably this could more than likely be false economy, as pointed out in (2) and (3) above.

There is a golden rule to cover all of the preceding statements and this can be defined as: *always use the largest package possible; unless space determines otherwise.*

A recent innovation, concerned with the use of plastic packaged parts, is the fully isolated TO-220 and TO-247/TO-218. This packaging requirement arises from the need to facilitate assembly on the part of some OEM manufacturers.

The argument which is put forward is that the lower overall mounting cost negates any premium which has to be paid for the single package. This could be true if the equipment manufacturer is producing in very large volumes. It should be pointed out that the higher thermal resistance of the fully isolated package could adversely affect reliability.

Other mounting considerations

Isolating washers

The original mica insulating washer, although being perfectly satisfactory in many respects, has apparently fallen from favour due mainly to the brittleness of the material and its inherent weakness and therefore propensity to be easily damaged by burrs, or other sharp edges and protrusions. The apparent fall from grace of the mica washer has led to the search for suitable alternatives.

Suitable alternatives are to be found from a diversity of materials. These range from beryllium oxide (BeO or beryllia) and other such ceramics to the anodised metals.

BeO, although possessing the lowest thermal resistance, is not favoured owing to its extreme toxicity and secondary hazard of being carcinogenic. It should be noted that the adverse publicity given to this toxicity is only valid for the dust of the material. A modicum of care should be exercised when handling the material and then it will be found to be relatively harmless.

Aluminium oxide (AlO_2 or alumina) has marginally inferior thermal resistance to beryllia but with none of the apparent toxicity of the former and, as such, constitutes a viable alternative.

The thermal resistance of these two ceramics is a function of the thickness of the material which in turn defines the 'insulation voltage' of the washer.

A third insulator is aluminium nitride. The use of this material tends not to be widespread since its use is only just being promulgated. The thermal resistance for this material is very nearly as good as that of beryllia and is certainly lower than that of alumina.

A fourth alternative is anodised aluminium. Unfortunately this material is characterised by having a higher thermal resistance than the two ceramics, and with poorer 'insulation voltage' capabilities. Its desirable feature is its lower purchase price. Against this should also be offset its inherent weakness to be damaged by burrs à la mica.

The use of the above insulators requires the use of *thermal grease* to compensate for the lack of surface finish which may be a property of the heat sink itself. This particular item is covered below.

The use of polymer based insulators is becoming increasingly popular and it is well to observe some precautions with such piece parts. The popularity of

5.3 OTHER THERMAL MANAGEMENT CONSIDERATIONS

these insulators stems from the advertised advantage of excluding the use of thermal grease. It has been my experience to note that some of these insulators do not stand up to compliance with the claim. I would therefore advise caution in the selection and use of such piece parts. The properties of these insulators should be exhaustively tested by the user to ensure that they do comply with the advertised claim.

The exhibited failure to comply with advertised claims is excessively high thermal resistance, especially if the use of 'grease' has been omitted.

Thermal grease

When mounting power package parts the use of one of the many proprietary thermal greases is strongly advised if the minimisation of thermal resistance is desired. This advice should be observed even if no insulating washer is employed. For proprietary reasons no recommendations can be made as to which particular brand of 'grease' is preferable. The reader is advised to be vigilant for some time to observe if any *drying-out* of the 'grease' is apparent. If drying-out is observed then the type of 'grease' should be changed.

In this respect some manufacturers greases are less prone to drying out than others.

Mounting clips

The use of screws to fix plastic packaged parts to heat sinks is not to be recommended. Excessive applied torque will cause deformation in the package resulting in either poor contact with the heat sink or worse, actual damage to the junction.

The TO-247 package is vastly superior than either the TO-218 or TO-220 in this respect. None the less failure to comply with the recommended mounting torque can easily lead to failure.

The use of mounting clips will overcome any tendency for failure to occur as a result of poor mounting owing to the effects of excessive torque.

Heat-sink considerations

All of the precautions outlined so far will prove to be of no avail if the flatness and quality of finish of the heat sink is found to be inadequate.

For lower power applications the heat-sink surface may be deemed to be adequate if the surface appears to be flat when viewed against a flat edge. This method of determining the merit of finish would be completely inadequate for high power applications. With high power usage the flatness, $\delta h/\mathrm{lm}$ (lateral measurement), if found to be better than 4 mils/inch, may be regarded as being eminently satisfactory. The surface finish is defined as the average value of the deviations above and below the average value of surface height. For satisfactory thermal performance a finish between 40 and 60 micro-inches may be regarded as being adequate. Obviously it would be desirable if the surface finish could be better than quoted, but the attainment of this finish should be judged against the higher cost to achieve the finer finish.

5.4 USING MULTIPLE JUNCTIONS (PARALLEL CONNECTION)

Discrete parts

One of the advantages of using power MOSFETS is the ease with which they can be parallel connected. This last statement is ostensibly true for the vast majority of conditions. The virtual sole exception to the rule is when the MOSFET has to operate within its unsaturated region of operation. Since linear operation at high power levels is the domain of the high power audio amplifier, the notes pertaining to this application will be included in Chapter 12.

The reason for parallel connection, in the first place, is to increase the current-handling capability of the power switch beyond the capabilities of a single junction — even if that junction happens to be the largest available. It also makes more sense to use two junctions which each run cold, than to use a single junction which runs 'hot'. If reliability is part of the cost equation, then the solution I have just discussed will be found to be more cost-effective, in the long term, than the use of a single MOSFET which will have a lower price tag.

The details associated with the connections and gate drive considerations for parallel connected parts will be covered more fully in Chapter 7 and will, therefore, not be discussed here.

MOSFET/IGT modules

If the use of modules, of either commercial or military type, is recommended to the customer the first reaction tends to be entirely predictable. The customer will usually multiply the price quoted for each discrete device by the number of dice within the module, then add some value for the cost of the module hardware and will then complain that the costs do not equate. This customer may tend to become suspicious and well imagine that he/she is being taken for the proverbial 'ride'.

No consideration is ever placed upon the fact that there is a price to be paid for convenience and the convenience is that the module is very much more easily mounted.

The convenience factor does not reflect the 'alleged greed' of the manufacturer but arises from the fact that the original cost equation is modified by such considerations as:

5.4 USING MULTIPLE JUNCTIONS (PARALLEL CONNECTION)

(1) *Yield.* When a module is manufactured, if only one of the dice fails the end result is failure of the entire module.
(2) *Labour.* Module manufacture may tend to be extremely labour intensive when compared with discrete part manufacture.
(3) *Improved thermal management.*

A glance at the data-sheet values for the thermal resistance of the virtual junction to case will demonstrate that the user will invariably have to mount more discrete devices than the module manufacturer to achieve similar values of thermal resistance, especially if insulation of the junction is required.

Taking all of this into account the module demonstrates that it possesses several worthwhile features.

When the module is mounted on the heat sink the user should endeavour to observe the same requirements that are observed when mounting discrete parts. The power dissipated by the MOSFET module is inevitably very much higher than any one discrete part and if the mounting procedures have tended to be 'sloppy' in the extreme then excessive junction temperatures and premature failure will be the inevitable consequence.

6
EMI/RFI and Layout

I am frequently asked to discuss all manner of techniques which I may recommend for the solution of various problems that I may encounter from time to time. Some of these techniques may be related to the reduction of elimination of both conducted and radiated interference. The answer that I invariably give is that it is always more desirable to eliminate the interference at its source than it is to filter out the interference after it has been generated.

It is frequently suggested by inexperienced users, that reduction of interference may well be some form of 'black art'. I am reluctant to admit that at one time I was also a subscriber to that particular philosophy. I should now say that nothing could be further from the truth.

EMI/RFI control is fundamentally the intelligent application of commonsense and also of the application of frequently forgotten theory. I am frequently amused at the possible inaccuracy of my last statement and its consequences, if black art it should prove to have become. Imagine the magicians who may be employed at the premises of those manufacturers of EMI/RFI filters. If their lore proved completely effective then the ultimate reasoning would be to have these people more gainfully employed in the design of the power circuits themselves.

I am dismayed at the lack of forethought given, by many engineers and technicians, to the subject of layout. Power electronic engineers should always be aware of the need to ensure that the layout of their circuit is not merely seen to be correct, but should, in reality, always *be* correct.

Aesthetics, although important, should take second place to correctness, regardless of the type of power switch which is used. The importance of correct layout takes on added significance with power MOSFETs — more so than for IGTs. It is frequently overlooked that the power MOSFET is a very high frequency amplifier and that its very speed, if not correctly employed, can in turn easily lead to the creation of interference.

This can best be explained quite simply. Consider the possible consequences of poor layout in a simple application as denoted in the layout of a *single diode injection laser* driven by a power MOSFET and illustrated in Figure 6.1. The injection diode load has no special significance in this instance. It was one of many loads that was employed among the many types of circuit with which I have been involved. The high peak current and short-duty cycle characteristic

88 EMI/RFI AND LAYOUT

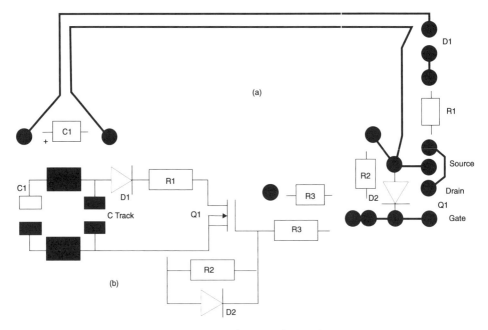

Figure 6.1 Poor PCB layout and equivalent circuit

of the load could easily have been illustrated by certain types of power converter circuit.

The poorly located decoupling capacitor C1 may initially be considered as having been compensated for by the close proximity of the positive and negative supply line tracks, thereby utilising the technique of *field cancellation*.

The equivalent circuit which is attached below the PCB part depicts the tracks from C1 to the *laser diode* D1 as being a transmission line which has not been correctly terminated. The Q of the circuit, the time element and the impedance characteristics of the line are not important in this instance. The vital point to be remembered is that the initial perturbation and possible reflections, which could be seen at D1 and possibly at the drain of Q1, may create the possibility of oscillations through unwanted feed-back. This prompts me to another observation which is specifically related to this subject.

I am intrigued at the implications of some of the questions that arise. These concern the probability in the mind of the questioner as to whether improvements in layout have any real bearing upon the reduction in generated RFI. It is absolutely essential for the first-time user to grasp this concept from the outset, since layout and EMI/RFI are inextricably linked. *Poor layout always increases the levels of interference.*

Having established this basic ground rule it is now desirable to examine the techniques which should be used to improve layout and to thereby reduce the emission of interference.

When examining the causes and effects of interference the reader should be aware that there are two types of interference involved:

(1) Conducted interference.
(2) Radiated interference.

It is extremely important to differentiate between the two types since their causes are not strictly related, and their effects and the methods used for their reduction, also tend to differ one from the other.

Conducted interference in general tends to be generated by rates of change of voltage (dv/dt), although this is not always the case, whereas radiated interference tends to be generated by rates of change of flux or current within conductors — di/dt. I would remind the reader that these are generalities, and that there are obvious exceptions to both rules. The exception to the radiated rule is the low pressure fluorescent lamp which acts as an antenna when used with an electronic ballast. The cause of the generated interference in this case is both i and v.

In the case of the fluorescent lamp the major cause of radiated interference is the rapidly changing electric field, due to dv/dt, across the length of the tube which acts as an antenna, with the secondary cause being di/dt resulting in fluctuations within the plasma and the density of the plasma inside the tube itself.

In the same application the causes of conducted interference is predominantly dv/dt which in turn cause rapid changes of current (di/dt) within the conductors; in this case the conductors are mainly the power lines. Note dv/dt is the cause and di/dt is the effect.

6.1 TECHNIQUES FOR REDUCING INTERFERENCE AT SOURCE

The source of interference largely depends upon the rapid transitions of voltage and/or current and is proportional to the rate of change and the magnitude of these transitions. Reducing the rapidity and size of the transitions is the only real method for achieving worthwhile degrees of reduction in the levels of interference. A variety of methods apply and these techniques can be reduced to the following four main sub-headings:

(1) Decoupling.
(2) Field cancellation and layout.
(3) Conductor length and circuit compactness.
(4) Topology (where applicable).

Decoupling

A careful examination of capacitors and their impedance characteristics, as a function of frequency, highlights the fact that no one type of capacitor can be regarded as being optimum for the purposes of decoupling power lines in circuits. The only satisfactory method to adopt in achieving near optimum decoupling is the use of *graded decoupling*.

Graded decoupling requires the use of various types of capacitor to cover the entire frequency spectrum of interest. This requirement extends to the use of an aluminium electrolytic type of capacitor for the lowest of frequencies, where the Equivalent Series Resistance (ESR) and Equivalent Series Inductance (ESL) tend to be of extreme significance. At lower middle frequencies solid tantalum electrolytic capacitors probably provide the optimum performance. The reader should be aware of the limitations in ripple current capability of the solid tantalum capacitor, if premature failure of the component is to be avoided. Upper middle frequencies require the use of plastic film capacitors of low inductance characteristics. A ceramic capacitor will be found to give the best cost/performance compromise at the highest frequencies. All of the decoupling capacitors should be connected in parallel and should be sited as near as possible to the part of the circuit requiring decoupling. *The quality of capacitors used for decoupling should never be compromised.*

If cost considerations must be taken into account then I would advocate the degree of grading. I will never support the possibility of utilising components of suspect quality as a means of economising over component selection.

Field cancellation and layout

Whenever an electric current flows through a conductor a magnetic field is set-up around the conductor in question. This should be remembered by the reader as being unequivocal. Equally any changes in magnitude to the current will result in a corresponding change in the magnetic field surrounding the conductor. Disregard for these two incontestable statements could lead to the generation of EMI/RFI.

Since it is impossible to eliminate the magnetic field surrounding a conductor without eradicating the flow of the electric current creating the magnetic field, the desired effect will be to maintain the field surrounding the conductor or conductors at a constant value. Initially this would appear to limit all currents within PCBs to be of the 'direct' variety.

If careful consideration is given to the problem it will be appreciated that the solution is more straightforward than was originally envisaged and was obliquely touched upon in the schematic of Figure 6.1. The solution is to group certain conductors in such a manner that the magnetic field surrounding these conductor groups should be held as near constant as possible. (The last statement, when

6.1 TECHNIQUES FOR REDUCING INTERFERENCE AT SOURCE

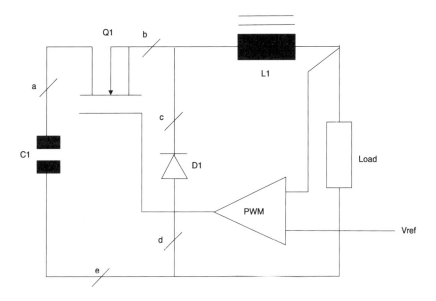

Note:
Only connector lines marked with a cross-hatch and a lower case letter
carry pulsating currents with no significant ripple current.

Figure 6.2 Schematic of the Buck converter showing sources of EMI/RFI

applied to power electronics, usually refers to a pair of conductors.) In reality it will not be possible to have power lines which only have a d.c. component of current in them. Some perturbations will always be present. The purpose of this part of the chapter is to see how it can be made possible that these perturbations are minimised.

In the Buck converter/chopper circuit of Figure 6.2 the points which are of most importance are marked with a lower-case letter to indicate where the nature of the currents are predominantly pulsating, with an additional ripple component. It should be appreciated that fields surrounding the current-carrying conductors will also be rapidly changing and will therefore become primary sources of interference.

If care is not exercised in reducing or cancelling the rapidly changing fields about these conductors, then the end result will be interference. The changes which should be made to the layout of the PCB which is recommended for this circuit are illustrated, by way of example, in the schematic of Figure 6.3. The reader should bear in mind that these recommendations are not the only techniques which may be used. I would advise the reader to pay a visit to the local reference library to obtain reading material which is devoted more or less exclusively to the subject.

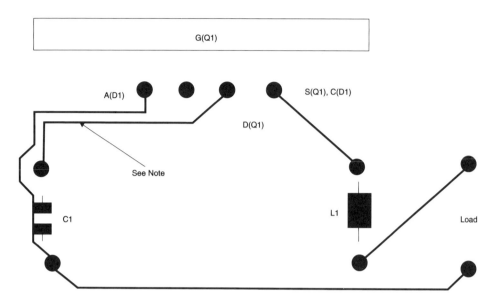

Note:
Maintain these tracks in close proximity to each other.

Figure 6.3 PCB layout of the power component part of Figure 6.1

The point of major interest is that conductors (a) and (c) have been placed in reasonably close proximity to each other. Since the magnitude of the currents flowing within these conductors is essentially the same, and because current is always present in one or other of the conductors, then the net value of the field which encloses the conductor pair should fundamentally remain constant.

The reader may feel that some liberties may have been taken with the packaging of Q1 and D1, by placing the two components within the same package. I would like to reassure the reader that this type of 'dual' component should be readily available and that if the preferred vendor fails to provide such a part, then an alternative supplier should be sought. The inclusion of the two semiconductors is not only expedient for thermal management, but encloses the two lines (b) and (c) for the purposes of 'field cancellation, whilst reducing the overall length of conductor (d).

It should also be noticed that no provision of output filter capacitance has been catered for. If this component should be required (to conform with specifications), then selection of this component will prove to be quite straightforward. For relatively large values of output current the usual method of determining the required degree of attenuation from reactances will be found to be inaccurate, owing to the fact that the capacitor may well be an electrolytic device. The method which is advocated does not include complexities and may be derived from:

$$V_{out(ripple)} = I_{o(ripple)}ESR(\text{output filter } C) \tag{6.1}$$

where the value of the output ripple is given in the overall specification, and the value for the ESR of the capacitor is obtained from data. It is the ESR of the capacitor that will determine the value of the capacitance, and the ripple current is derived from:

$$\delta i_{out(ripple)} = [(E_{in} - E_{out})t_{on}]/L1 \tag{6.2}$$

where the ripple current $i_{out(ripple)}$ is expressed as a peak-to-peak value and the r.m.s. value $I_{o(ripple)}$ is approximately 10% of the average value of the output current I_o. t_{on} is the ratio of duty cycle D and switch frequency f_{sw}.

In order that $L1$ be defined the value of $i_{out(ripple)}$ should be determined initially. $i_{out(ripple)}$ may also be expressed as:

$$i_{out(ripple)} = (3)^{1/3}I_o/10 \tag{6.3}$$

Equating (6.2) and (6.3) and substituting for duty cycle and switch frequency yields:

$$[(E_{out}/f_{sw})(1 - D)]/L1 = (3)^{1/3}I_o/10$$

But E_{out}/I_o is equal to the load Z_{load}. Therefore:

$$L1 = [10Z_{load}(1 - D)]/[(3)^{1/3}f_{sw}] \tag{6.4}$$

The solution of (6.4) would apparently require that E_{in} be static. Solution of $L1$ should not prove to be a complication. $L1$ can be calculated for a single value of E_{in}, which may be obtained from the input specification. The pulse-width modulator (PWM) will compensate automatically for any reasonable value of E_{in} greater than E_{out}.

Conductor length and circuit compactness

Where field cancellation cannot be accommodated without extreme difficulty conductor lengths should be maintained as short as possible, in order that $L\, di/dt$ effects be minimised. The optimum route for achieving circuit compactness will be by the miniaturisation of complete circuits either by *monolithic integration* or *hybridisation*. If this type of miniaturisation is not economically acceptable (or feasible) then the use of *surface mount* technology should be examined. If the use of surface mount devices (SMD) is also not possible then the only recourse is to maximise the packing density of printed-circuit boards (PCB).

94 EMI/RFI AND LAYOUT

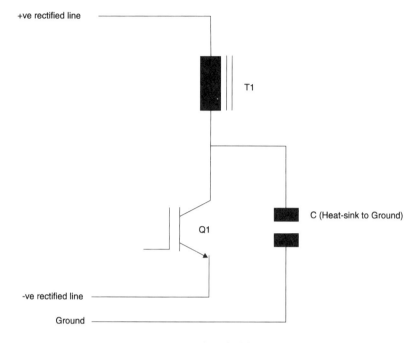

Figure 6.4 Convecter with unshielded stray capacitance

Investment in Computer Aided Design (CAD), along with Computer Aided Manufacture (CAM), in order to facilitate increased packing density should be investigated. This type of investment should not be considered as being prohibitively expensive. Inexpensive software for IBM PCs (and compatibles) is readily available.

Multi-layer PCBs, although being a prerequisite for maximum packing density, should be used with caution by the inexperienced user. A lack of care in its use could introduce unwanted stray capacitances which may create undesirable displacement currents — which could in turn lead to the generation of interference.

An important part of this sub-heading is the use of capacitively insulated shielding and its judicious use in the elimination/reduction of interference which may be induced by certain packages into the ground line. The coupling of this interference is mainly created by the the existence of stray capacitances.

Consider the schematic of the simplified power converter of Figure 6.4.

If the transistor Q1 were to be mounted on a grounded heat sink, and this is especially important in certain instances, namely that of off-line applications, then some form of insulator must be included between transistor case and ground. This introduces the capacitance C_{stray} as shown.

6.1 TECHNIQUES FOR REDUCING INTERFERENCE AT SOURCE

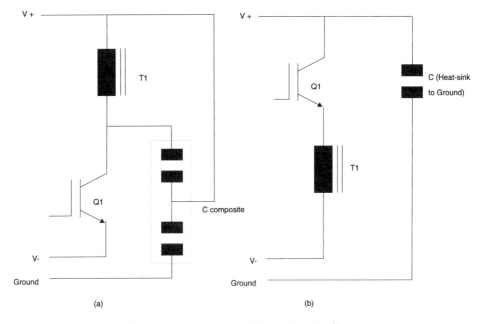

Figure 6.5 Relocation of C_{stray} and or its effects

This value of capacitance will vary between 2 pF and 10 pF depending upon the package. If the transistor Q1 were to have voltage rise and fall times (t_{rv} and t_{fv}) of 100 ns; then the maximum dv/dt at the collector/drain of Q1 would be 3730 V/μs (in a UK off-line circuit). This would create interference currents (through C_{stray}) which will range in value from 7 mA to 37 mA, again depending upon the package of Q1. These levels of interference currents are undesirable to say the least. If faster switching of Q1 were to be initiated then the level of interference current flow through C_{stray} would be untenable.

Two readily available solutions may be implemented in order to reduce or eradicate the displacement current through C_{stray}. These two solutions are given in Figure 6.5.

The solution of Figure 6.5(a) invokes the use of shielding whereby a conductive layer, of copper, is sandwiched between two isolation layers. This sandwiched conductive layer effectively creates a second capacitor in series with the original C_{stray}. The significance of this second capacitor is the connection between the two capacitors to one or other of the raw d.c. bus lines (rectified a.c. line input). This means that the displacement current due to the dv/dt of 3730 V/μs is routed into the d.c. bus line while the interference current into the ground line is created by the 100/120 Hz ripple voltage component. The peak dv/dt could be found to be reduced to 40 V/μs. The reduction in interference current is therefore approximately equal to five orders of magnitude.

The solution of Figure 6.5(b) is a much more radical and requires a little forethought. Moving the position of Q1 to the *high-side* of the transformer primary results in the displacement current through C_{stray} to be the same as that of the lower value of current — obtained for Figure 6.5(a) — but without the use of shielded insulators.

Either of the two techniques may be used with equal success, the choice could well be the result of personal preference on the part of the reader.

Magnetic components and related capacitances

One of the problems which has been discussed is the stray capacitance associated with transistor packages. Another stray capacitance which is frequently overlooked or possibly neglected is the capacitance, between windings and the core, within transformers and inductors. Interwinding capacitances may also be placed into this category. The effects of these capacitances are mainly manifest when the cores are grounded to the chassis. Discussions related to the capacitances of magnetic components can be sub-divided into two categories:

(1) *Transformers.* In off-line power supplies it is considered as being quite normal (for safety legislative conformity) for the transformer to contain a safety electrostatic screen between the primary and secondary and for this safety screen to be grounded. Perturbations on either the primary or secondary windings will be coupled into the ground line as conducted interference. The method most accepted for reducing the capacitive effects of this safety screen is the emplacement of further electrostatic screens between the windings and the original safety screen, with these screens being connected to some d.c. potential, thereby reducing the likelihood of injecting interference currents into the ground conductor.

(2) *Inductors.* In switched-mode supplies inductors are usually employed as averaging transducers. Seldom is there any thought given to the capacitance between the winding and the core. Even less frequently is there any consideration given as to which end of the winding will have the most capacitance to the core. If the end of the winding with the most capacitance should be connected to that part of the circuit having the greatest signal excursions it will become obvious that the result could be more emissions than was originally allowed for. The solution is simple in the extreme and involves nothing more than the reversal of the connections to the inductor.

Stray capacitances associated with capacitor cases

Large aluminium electrolytic capacitors are frequently used as reservoir capacitors in high-powered supplies. Many readers may well be unaware that the *cathode* (the negative plate) is in electrical contact with the case of the capacitor. It is normal for these large capacitors to be clamped in close proximity with the chassis (which is usually at ground potential). This proximity between case and chassis constitutes yet another capacitance, and perturbations on the negative supply rail will induce a displacement current through this stray capacitance. If the case of the large electrolytic is suspected of being the source of interference then the remedy is to institute shielding techniques. Insulated aluminium foil should be wrapped about the capacitor and connection of this foil to some innocuous part of the circuit should be carried out experimentally.

Topology considerations (where applicable)

If the converter of Figure 6.4 was of the *forward* variety then the current through Q1 and the transformer primary would be pulsating in character — and could require the adoption of all of the other sub-headed techniques for reducing the possible interference at its source.

On the other hand, if the converter should happen to be of the *fly-back* variety (and its mode of operation was characterised as continuous), then the current through the transformer would also be continuous, but superimposed by a ripple component. (Continuity of current is implied as being through the primary and the de-magnetising windings alternatively.) The interference created by the fly-back converter, operating in the continuous mode, can be demonstrated to be significantly easier to reduce at source. The simple change in converter topology (along with operational mode), and the profound effects upon possible interference, infers that the chosen topology can influence the original interference level and the subsequent ease or otherwise in the reduction of interference.

Certain topologies will be found to be demonstrably better than others at requiring less interference reduction. The most notable of these topology varieties is the family termed 'cascaded'. The optimum of these cascaded topologies is that of the *boost cascaded by a Buck*. This variety of cascaded topology is frequently referred to as *coupled inductor converters.*

One of the major attributes of coupled inductor converters is that, by judicious adjustment of the ratio of input and output inductances and also by adjustment of the degree of the coupling, the converter can be adjusted to have almost zero ripple characteristics on either input or output. This zero ripple attribute should be considered when the choice of converter topology has to be decided.

The reader is encouraged to read literature or textbooks related to the advantages and disadvantages of converter topologies and to their design. This subject is beyond the scope of this particular volume.

Some of the material contained within this chapter has very little connection with MOS-gated transistors; however, if the material should prove to be of interest to the reader then its inclusion will be of value.

7
General Circuit Techniques

During the many conferences which I am, on occasion, privileged to attend and also at the various seminars that I have presented in the past and do still present, I am frequently left at a complete 'loss for words' at some of the questions which are raised. The questioners are not confined to any one age group in particular but extend over the complete age spectrum. The questions themselves reflect, in some instances, a complete misunderstanding of fundamentals.

It is also my experience that no one country's inhabitants can be singled-out as beig more in the vanguard of this regrettable situation. It may well be a reflection upon all of mankind. It is certainly not my intention to try and set to rights the apparent lack of training. In this chapter I will endeavour to impart, to the reader, the many techniques that I have been exceedingly fortunate to have learned over the years.

Much of this chapter may at first appear to have been considered in a very haphazard fashion and this may well be true to some extent, although it has not been my intention to present it in such a manner.

7.1 USING MEDIUM VOLTAGE MOSFETs IN HIGH VOLTAGE CIRCUITS

When working off of rectified 3-phase mains, it is not always essential to use MOSFETs with a BV_{dss} equal to or greater than the value of the raw d.c. rail which usually indicates the value of the peak rectified line voltage. This statement will hold true equally well for IGTs. It is pertinent to substantiate this claim, at this juncture, since failure to do so would invalidate the claim itself.

In order to substantiate this claim it is necessary to explain that the load should, in my experience, be the primary winding of a transformer. I have been advised on several occasions that this is not always the case. Upon reflection I return to my original assumption. If it can be proven that my original supposition as to the type of load is erroneous then I will withdraw from this assumption. I would like to point out to the astute reader that I am aware of the use of series connected silicon controlled rectifiers (SCRs) in certain high voltage traction applications. Towards this end I will illustrate, later in the chapter, how MGTs

100   GENERAL CIRCUIT TECHNIQUES

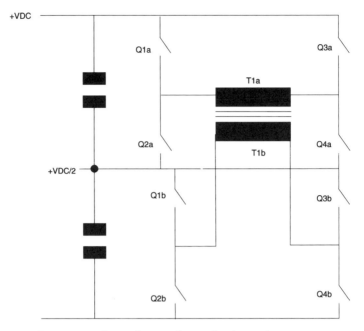

Figure 7.1   Circuit showing the use of 'split supply' connection

may be series connected in order that they may work in applications where the load is not a transformer. It is an application that I do not recommend, but merely include as a point of interest.

Consider the circuit in Figure 7.1, where the full-bridge inverter is loaded by a transformer. Let V_{DC} be derived as follows:

$$V_{a.c.in} > 380 \text{ V}$$

Therefore, $V_{d.c.}$ will be not less than $2^{0.5} \times 380$. This value of $V_{d.c.}$ will therefore not be less than 530 V.

If $V_{a.c.in}$ is not greater than 550 V then the individual MOSFETs in the circuit (Q_1 to Q_8), should have a minimum BV_{dss} of 500 V and no more.

The operation of the circuit is conventional with switches Q1 (a and b) and Q4 (a and b) diagonally conducting during one part of the a.c. cycle and Q2 (a and b) and Q3 (a and b) conducting during the remaining part of the cycle.

When Q1a and Q4a are 'on' the left hand side of T1a (primary) is connected to $+V_{d.c.}$ while the right hand side of the transformer primary is connected to $+V_{d.c.}/2$.

When Q2a and Q3a are 'on' the right hand side of T1a (primary) is now connected to $+V_{d.c.}$ with the left hand side being connected to $+V_{d.c.}/2$.

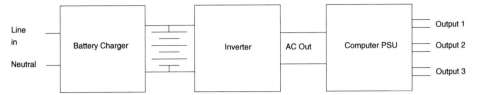

Figure 7.2 Complete connection sequence of UPS and computer

Primary winding T1b will be switched in synchronism by Q1b and Q4b and by Q2b and Q3b. Particular attention should be paid to ensuring that the starts and the finishes of both primary windings are adjacent to each other.

Provided the conduction intervals are the same, when the diagonally opposing switch pairs are closed then the volt-second saturation integrity of the transformer core has not been found to have been violated and my original claim has been substantiated.

In case the reader is concerned over the likelihood of imbalance between the potential differences of the two primaries, then let me offer the assurance that the supply-half which is greater in magnitude will in reality supply the greater primary current until voltage balance is again established. This self-balancing tendency is one of the features of this type of circuit configuration. The same claim can be demonstrated to be valid for other converter topologies.

7.2 UNINTERRUPTIBLE POWER SUPPLIES

I am intrigued at the evolutionary path that this particular type of power supply has undergone. It is obvious that there is a need for such a type of supply and the need becomes evermore evident with the frequency of occurrence of *brown-outs*, etc.

It is not the need for uninterruptible capability which is in question, merely the execution of the fulfilment of the need. Most readers will be aware of the need for the UPS, especially those readers who are in frequent contact with computers — be they the humble PC or a large mainframe. In terms of overall efficiency it is difficult to envisage a worse configuration than that of Figure 7.2.

Efficiency here covers all aspects — such as power, use of materials and resources and also includes cost.

Several analogies relating to the inefficiency and to the inappropriate use of materials, etc., spring to mind, concerning present-day configurations, but none of these analogies appropriately demonstrate the incorrect solution arrived at by evolutionary processes. Illustration of the point concerning overall efficiency is simplicity itself. This may be expressed as:

$$\text{System efficiency} = UPS_{\text{Efficiency}} \times PSU_{\text{Efficiency}} \tag{7.1}$$

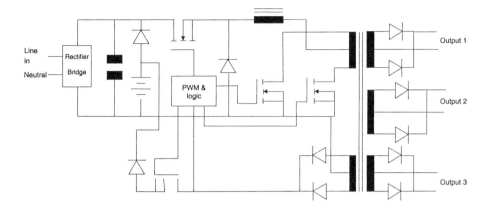

Figure 7.3 Schematic of uninterruptible off-line SMPS

where $UPS_{Efficiency}$ is the efficiency of the UPS and $PSU_{Efficiency}$ is the efficiency of the power supply inside the computer itself.

By way of example: if the efficiency of the UPS and the power supply of the computer were to be given the same efficiency of, say, 80% then the system efficiency is found to be an unimpressive 64%.

The more efficient solution to the requirement would be to make the computer power supply uninterruptible in its own right. An example of such a power supply is illustrated in Figure 7.3.

The circuit in Figure 7.3 is not meant to indicate the usage of medium voltage MGTs in a high voltage application. It is merely included to demonstrate the principle of system efficiency improvement.

The power supply in Figure 7.3 can easily be shown to have an overall efficiency of 80% or greater, which should make this type of supply more attractive. Of more significance is the greatly reduced system cost owing to the virtual elimination of the UPS in Figure 7.2.

The principle which I am advocating, in this instance, is not to accept earlier solutions as being optimum. Its evolution could well have been the result of a very differing set of circumstances. It therefore behooves the reader to look at today's problem in a fresh light.

7.3 EMITTER SWITCHING WITH BIPOLAR JUNCTION TRANSISTORS

The use of transistors having a blocking ability of 600 V or greater is frequently a requirement in *single-ended off-line converters* which may be required to operate at several hundred kilohertz. The MOSFET of 800–1000 V BV_{dss} is one obvious

7.3 EMITTER SWITCHING WITH BIPOLAR JUNCTION TRANSISTORS 103

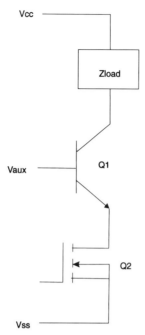

Figure 7.4 Schematic of the Bi-MOS cascoded compound switch

choice for this particular application. Its relatively high on-state resistance is frequently considered as being a serious drawback. We will therefore explore another possibility.

At first sight it would be easy to assume that this section is irrelevant, since the book is supposed to be about MOS-gated transistors. As the reader progresses through the section it will become more apparent that the section is completely relevant.

BJTs suffer from two important defects which to some extent limit their overall desirability. These limitations may be stated as being:

(1) A complete lack of ability to switch extremely rapidly.
(2) The absence of a well defined safe operating area (SOA) in both the forward-biased and reverse-biased states.

Both of these limitations are overcome to a greater extent by operation in *common base*. Common base operation is easily achieved by the use of the technique known as *emitter switching*. It can also be demonstrated that this technique is most easily achieved by the use of a modestly sized low voltage MOSFET connected in a cascode connection with the high voltage BJT. To be fair to other proponents of emitter switching the technique shown in Figure 7.4 is not the only way in which cascode switching may be achieved.

104   GENERAL CIRCUIT TECHNIQUES

In the circuit depicted in Figure 7.4 the cascode connected BJT and MOSFET pair are found to donate almost all of their strong points to the overall circuit, whilst not permitting the transfer of virtually a single disadvantage.

The BJT is now seen to have a superior RBSOA, and to provide the switch with an overall blocking voltage equal to V_{CBO}.

The speed of the combined switch would appear to be that of a MOSFET, and the switch is capable of operating in switch mode supplies with frequencies of several hundred kilohertz. The operation of the circuit is found to be quite straightforward with few, if any, hidden pitfalls. A brief description of the operation is as follows:

(1) Initially assume the switch to be off, with the collector of Q1 being at V_{cc}, the base at V_{aux} and the emitter at some voltage (which is not sharply defined) between V_{aux} and V_{ss}. The drain of Q2 is at the same potential as the emitter of Q1.
(2) The application of the positive going edge of the gate waveform at Q2 causes Q2 to turn on, pulling the drain of Q2 and the emitter of Q1 towards V_{ss}. This action causes charge to be transferred rapidly from C1 into the base of Q1, thereby causing Q1 to rapidly turn on. Base current for Q1 is now supplied from V_{aux} and maintains Q1 in its on state.
(3) At the termination of the gate drive pulse to Q2 — when the gate of Q2 is taken to V_{ss} — Q2 is turned off and its drain attempts an excursion towards V_{cc}. This positive excursion of the drain of Q1 and the emitter of Q2 is clamped by the base–emitter junction of Q1 (now in a state of avalanche), and the full value of collector current I_c rapidly removes the stored charge in the base–emitter junction of Q1. This method of extracting the charge from the base of Q1 is found to be relatively benign in nature, without the tendency of current crowding in Q1 occurring.
(4) When all of the excess charge has been removed from the base of Q1 the collector current now transits into the fall-time phase, which in turn corresponds to the recovery time of a very fast diode. The consequential fall-time of I_c is found to be surprisingly short.

The switching behaviour of the compound switch (as described in (1) to (4) above) results in very low switching losses in a switch with high voltage blocking capability. The switching losses more than compensate for the added conduction losses of the MOSFET Q2 and the total losses of the combined switch will be found to compare extremely favourably with a high voltage MOSFET — even at frequencies of several hundred kilohertz.

The selection of Q2, between BJT and MOSFET, will be found to be overwhelming in its favour towards the MOSFET, on the grounds of vastly superior performance of the switch, with the added feature of the ease of drive of the MOSFET.

Component selection for the circuit of Figure 7.4 is also found to be comparatively simple with the exception of C1. In this instance it is found that this particular component tends to result in a compromise value being chosen, since its value can affect both turn-on and turn-off of the overall switch. Too large a value results in extremely rapid turn-on with an equally rapid drop into *hard saturation*. Unfortunately this turn-on speed is found to be achieved at the expense of rather leisurely performance at turn-off. Too small a value for C1 results in slower turn-on speed, but with very rapid turn-off capability both in storage time and fall time.

As stated previously the schematic in Figure 7.4 illustrates only one of several ways of achieving cascode connection of the BJT. The added losses of the series connected low voltage MOSFET can be negated to some extent by connecting a current transformer in series with the emitter of the BJT and have the switching performed on the secondary side of the current transformer.

7.4 SERIES CONNECTION OF MGTs AND SLAVE CONTROL OF THEIR GATES

This section may be regarded as an adjunct to earlier sections, namely, 'Using medium voltage MOSFETs in high voltage circuits' and 'Emitter switching with bipolar junction transistors'.

In the circuit of Figure 7.5, the MGTs Q1 and Q2 are both 400 V devices and are included in order to fully illustrate the principle.

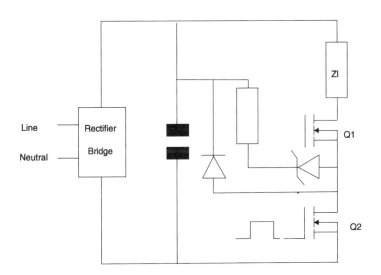

Figure 7.5 Series connected and slaved MGTs

It should be understood by the reader that a single 800 V MGT would have furnished a more cost-effective solution. MGTs of 400 V are eminently suitable in 220/240 V off-line applications since the drain of Q2 is clamped to the raw d.c. bus and it is obvious that Q2 will never be exposed to more than 400 V, thus enabling Q1 to perform the primary task of holding off the remaining voltage which could be generated by the leakage inductance of the transformer.

7.5 ULTIMATE SWITCH SPEED, ITS RAMIFICATIONS AND ATTAINMENT

This section does not involve the recommendations of observing good layout in order to reduce emissions of unwanted interference, although the importance of this precaution can never be over-stated. The section deals with techniques which should be considered if the ultimate speed of switching is to be attained. High switching speeds are a prerequisite in very high frequency operation if losses are to be minimised.

The requirements which should be observed, if high switching speeds are to be attained, are diverse and encompass such titles as: switch transistor packages, layout, circuit strays drive circuits, amongst several topics. A brief discussion of most of the sub-headings will be considered in this section.

Packaging of switch transistors is one area in which the semiconductor manufacturer can play a very significant role in the achievement of optimum performance. The reader should be aware of the serious limitations of those old favourites the TO-3 and TO-66, so beloved of the military. Several aspects of these particular packages make them totally unsuited for really high switch speeds.

The flange of these two packages are usually fabricated from steel for reasons of cost. The glass to metal seals of the lead-outs of gate/base and source/emitter are the creators of significant values of undesirable inductance which swamp the effects of lead inductances. The flange itself as being the drain/collector connection also introduces inductances in the drain/collector circuits although the effects of these inductances are less profound.

Plastic power packages (TO-220, TO-218 and TO-247 and their metal hermetic equivalents) go some way in redressing some of the deficiencies which have been covered, but unfortunately do not address the deficiency of the power source lead inductance and the negative feed-back upon achieving the desired speed. Multi-lead plastic packages as used in the manufacture of the current mirror MOSFET probably achieve the best possible compromise which is affordable. These packages have an auxiliary source connection for the gate return signal which eliminates the $L\, di/dt$ effects of the power source connection.

Junction size can play a significant role in achieving the ultimate switch speed in a circuit. Large junctions require large spikes of current to be delivered in a short period of time to the gate (input capacitance) in order that the part can

7.5 ULTIMATE SWITCH SPEED, ITS RAMIFICATIONS AND ATTAINMENT

be made to switch rapidly. Gate drive circuits become ever more expensive as their source and sink capability is increased. I am not at this stage advocating driving two smaller junctions in parallel (as opposed to one large junction), since the improvement in speed which may be achieved by this approach can be minimal. The technique which I propose is to *cascode connect two MOSFETs*. This can be accomplished by simply modifying the circuit of Figure 7.4.

To illustrate the principle consider driving one of International Rectifier's IRFP460 MOSFETs (or its true equivalent). This device has a C_{ISS} equal to 4100 pF, requires a gate charge Q_G of 120 nC and has an $R_{ds[on]}$ of 0.24 Ω. Using the gate charge curve from the data sheet it soon becomes apparent that a charge of 75 nC is required to transit the part from the cessation of $t_{d(on)}$ to the end of t_{fv} the voltage fall time. To achieve a transit time of 25 ns would require a peak gate current of 3 A.

If the modified circuit of Figure 7.4 is used the driven MOSFET could be an IRFZ44. The total switch $R_{ds[on]}$ will increase to 0.268 Ω (an increase of 11.67%), but the driven C_{ISS} is reduced 2500 pF (a reduction of 39%) and the required gate charge Q_G reduces to 69 nC (a reduction of 42.5%). Once again using the gate charge curve, but not it is the one for the IRFZ44, we find that the charge required for the same transition is a mere 39 nC. The same transit time of 25 ns now requires 1.56 A (a saving of 48%).

A further expansion of the idea would be the parallel connection of two IRFP450s with the IRFZ44 being replaced by an IRFZ34. Conduction losses would be marginally better all round than for the compound IRFP460/IRFZ44 switch, while the transit time of 25 ns would require a peak gate current of 1.04 A.

Few additional precautions, if any, are required over and above those shown in the circuit of Figure 7.4, except the replacement of the 5 V Zener in the base circuit of the BJT with one of 10 V for the gate of the large MOSFET.

The lesson of the above is that the occasional diversion from conventional thinking can reap unexpected rewards.

Layout and this includes packaging, but not in the sense that has been considered in the immediately preceding paragraphs, is another topic of discussion which should be considered carefully if the switch speed is to be maximised in an optimum manner. In this context the keeping of compact dimensions is a prerequisite in minimising circuit strays and their effect upon speed. Hybridisation is perhaps the major method of achieving compact size while overcoming the effects of the power switches package at one and the same time.

When considering the gate-drive circuit, especially for a high-side switch, the reader should be aware of the potential pitfalls which could be the consequence of utilising poorly designed and/or constructed drive transformers. Leakage inductance of these components, if not minimised, can have a deleterious effect upon the achieved speed. One alternative to this component is the use of specialised gate driver integrated circuits which remove the necessity for drive transformers, whilst enabling the achievement of compact drive circuits.

The reader should be aware by now that there is a considerable degree of overlap in the attainment of high-switching speed and the reduction of interference emissions. This chapter only serves to reiterate an earlier observation.

7.6 USING STANDARD MOSFETs IN HIGH FREQUENCY PWM INVERTERS

The parasitic *body–drain diode* within standard MOSFETs is not renowned for having a short reverse recovery time, after having been under forward-biased operation. This same parasitic fortunately has a certain degree of ruggedness so that early failures owing to dv/dt application during recovery are a thing of the past. Today's failures tend to stem from excessive dissipation due to recovery time losses.

Certain manufacturers have attempted to overcome this lack of speed by the judicious use of life-time control of the minority carriers. The use of life-time killing certainly does have the effect of reducing t_{rr}, but at the expense of increases in $R_{ds[on]}$ and conduction losses. The improvement in t_{rr} also tends to be temperature dependent, to a greater or lesser extent, depending on the type of dopant used for life-time control.

The techniques which were advocated (to forestall dv/dt inducing failure) have been widely documented and included the following recommendations:

(1) Total isolation of the body–drain diode by connecting a Schottky diode in series with the MOSFET and to then by-pass the series connected pair by a fast anti-parallel diode connected externally.
(2) Active or passive control of the applied dv/dt across the diode during reverse recovery.
(3) Current fed inverters.

All of the foregoing recommendations have been found to be successful but have very little relevance for present-day conditions. The objective which should be considered as being of almost paramount importance when using MOSFETs with standard but rugged parasitic body–drain diode is the limitation of peak reverse current I_{RM} during diode reverse recovery. This requirement arises out of the need to maintain recovery losses to an acceptable level and thereby minimise total power dissipation within the MOSFET. This minimisation of losses has the twofold effect of enhancing the reliability of the circuit overall whilst ensuring that the resultant junction temperature is maintained as low as possible, thereby further reducing the operating $R_{ds[on]}$ with a consequential

7.6 USING STANDARD MOSFETs IN HIGH FREQUENCY PWM INVERTERS

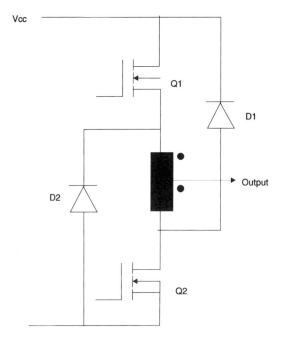

Figure 7.6 Circuit for PWM inverters using standard MOSFETs

reduction in overall dissipation. The loop can be seen to be both self-enhancing and, within reason, to be reiterative.

The circuit in Figure 7.6 demonstrates just one technique for limiting I_{RM} through the body–drain diode during the reverse recovery phenomenon of the diode in a PWM inverter. The physical size of the auto-transformer T1 is modest since limited volt-second saturation capability is required. Modest current handling ability is an attribute that also applies to the external anti-parallel diodes D1 and D2. The need for modest current diodes is due to the parasitic diodes of the respective MOSFET conducting the bulk of the commutated current. The diodes D1 and D2 are only necessary during the time that the auto-transformer core transits from one level of B_{sat} to the other level of B_{sat}.

The method used for determining the core requirements for T1 is as follows:

(1) Using the following equations:

$$V_{cc}t_{rr} = NA_e\delta B_{max} \qquad (7.2)$$

$$L = V_{cc}t_{rr}/I_{RM} \qquad (7.3)$$

$$N = (L/A_L)^{0.5} \qquad (7.4)$$

where V_{cc} is the maximum value of the raw d.c. bus, t_{rr} is the maximum reverse recovery time of the body–drain diode (from data sheets), A_e is the area of the core under consideration, $\delta\beta_{max}$ is peak-to-peak value of flux density from $-\beta_{sat}$ to $+\beta_{sat}$, the ratio of t_{rr} and I_{RM} provides both the maximum recommended di_F/dt (for the diode) and the peak diode current for the external anti-parallel diode and A_L is the core inductance factor.

Calculate the number of turns required to provide the necessary inductance in order to limit both the di/dt and the peak I_{RM} through the diode.
(2) If all of the criteria are not satisfied with the core under scrutiny repeat the process for a marginally larger core.

It should be remembered that the selection of the core of T1 by maintaining its size to as small as possible is vital if both the cost and the overall size is to be optimised.

The points raised over the use of the standard MOSFET's body–drain diode is meant for applications where the frequency of switching is not considered to be excessive, namely 20 to 50 kHz. In Chapter 12 I propose to demonstrate applications where the use of this parasitic is not really possible, even when life-time control of the parasitic has been employed during fabrication.

One further point that the reader should be aware of is the necessity to maintain turn-off gate circuit impedance to as low a value as is possible. I have already covered this subject previously related to the phenomenon of Miller turn-on. The unfortunate consequences of not paying heed to the previous advice cannot be over-stated.

7.7 HIGH FREQUENCY OPERATION OF IGTs

Second generation IGTs from certain manufacturers hold out the potential for high frequency operation to several hundred kilohertz. It should be pointed out that this refers to a *quasi-resonant mode* of operation. Alternatively non-resonant operation is possible up to 50 kHz provided the collector current is suitably derated. One device which has recently been introduced permits *quasi-square-wave* switching up to 50 kHz without any derating in collector current. Unfortunately, 50 kHz is normally not recognised as true high frequency, especially where SMPS are concerned, high frequency qualification usually implies 200 kHz and upwards. Operation at these frequencies is possible provided some form of *switching aid* is incorporated. These switching aids could be snubbers which apply very heavy snubbing to the voltage rise time. The drawback to this form of switching aid is the severe losses which must be incurred in the snubber resistor if one is used, or the complexity of a high current non-dissipative snubber which must be resorted to.

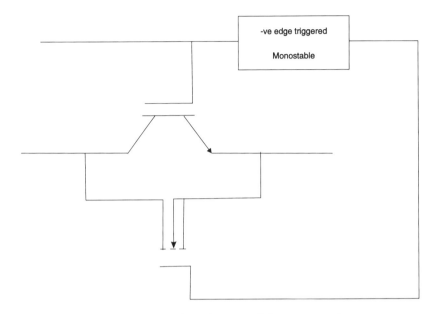

Figure 7.7 IGT and MOSFET in high-frequency switching

7.8 MOSFET SWITCHING AID TO IGT FOR HF SWITCHING

The interesting alternative to this is the parallel connection of a MOSFET across the IGT. The concept for this connection has been promoted in the past as one means of driving BJTs, where the MOSFET controls the switching while the BJT handles the conduction interval.

The overall circuit for this application is given in Figure 7.7.

The monostable is triggered-on by the trailing edge of the IGT gate pulse. The duration of the on-time of the monostable equates to the total switch-off time of the IGT. After that the MOSFET may be driven off as rapidly as possible in order to minimise the overall losses.

It can be demonstrated that in the application illustrated in Figure 7.7 an IGT can be used at the full I_c rating at 250 kHz with the addition of a relatively small MOSFET. Any other two-package combination would not perform with the same overall efficiency. Efficiency in this instance includes the relative cost of the complete solution.

7.9 PARALLEL CONNECTION OF MGTs

Most of the information here is related to the medium and high voltage MOSFET, since discrete IGTs usually tend to be high current devices. This is a generalisation

112 GENERAL CIRCUIT TECHNIQUES

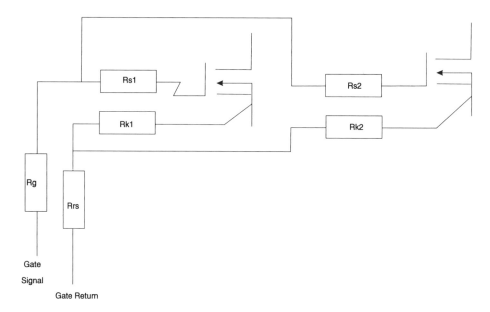

Figure 7.8 Gate connection details of a parrallel pair of MOSFETS

and does not cover all circumstances. Parallel connections of MOSFETs until recently tended to be surrounded by a certain air of mystique. It was considered mandatory that circuit layout symmetry was a fundamental requirement if good current sharing was to be achieved. I would like to add that symmetry of layout is desirable but not essential.

In the schematic of Figure 7.8 the niceties of drive requirements to avoid dynamic unbalance in the two parallel connected MOSFETs are displayed. The resistors R_S and R_G are to ensure that the slower switching device will draw more current from the gate-drive circuit and thereby slow down the faster part. This will become obvious if the gate-charge curve is also considered at this point.

The slower switching part will have a gate voltage that is lower than its much faster sibling. Ohm's law will dictate that the slower part will therefore have more current forced into its gate terminal during turn-on and thereby its turn-on speed will be enhanced. The converse is true during turn-off. The resistors R_K and R_{RS} decouple the power sources and the power current path from the return of the gate-drive circuit. It is essential that the resistors R_{K1} and R_{K2} be connected directly to the source connections of their respective MOSFETs. Static unbalance will be taken care of by the device with higher t_j handling less current by virtue of its higher $R_{ds[on]}$, always assuming that the two devices started with similar values of $R_{ds[on]}$ at $t_{case} = 25$ °C.

It is vital that the *gate-drive circuit ground* is connected at the negative terminal of the voltage source of the main power circuit. Failure to observe this recommendation could result in mistriggering of one or other of the two MOSFETs.

7.9 PARALLEL CONNECTION OF MGTs

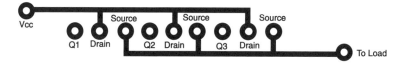

Figure 7.9 Symmetry of power circuit

The need (or otherwise) for symmetry in the power circuit, and its nuances is covered in the schematic of Figure 7.9. The diagram represents the upper trackside of a printed circuit board. It should be noted that the *common drain connection* for the power circuit is made at the drain of Q1 while the *common source connection* is taken from the source of Q3. The explanation for the method adopted here is as follows: any tendency for Q3 to switch faster than either of its two companions, owing to the reduction of negative feed-back from power circuit source inductance, is compensated for by the combined drain circuit inductance acting as a turn-on snubber, thereby reducing the rise in drain current in Q3. Similarly, Q1 which should have the greatest amount of negative feed-back has the smallest turn-on snubber. The net result of the actions described is a tendency for dynamic sharing to take place.

At this point no effort has been made to balance the power circuit or to consider any of the ramifications which would arise out of any imbalance.

The reader should note that the concept in Figure 7.9 is totally at odds with earlier thinking which required individual drain connections to the common drain connection and likewise for the individual source connections to the common source connection to be matched as much as possible for length.

8
Power Supplies

Many times I have been asked how and why I became involved with the subject of power supplies. The following has been frequently added to the question for good measure: *'After all, any fool can design a power supply'*.

The offensive nature of such a quotation merely indicates the ignorance with which the humble power supply is perceived by some individuals. It is possible for anyone, with the minimum of both training and knowledge of the subject, to design a simple linear power supply in a relatively short time-frame. What is more is the relatively good performance which may be achieved by this beginner's first attempt. On the other hand, I would welcome the opportunity to observe how the said 'any fool' would attempt the design and construction of a reliable switching regulator which performed moderately well. This item is a system in its own right.

It is certainly not my intention to go into the details of the design of any type of supply, since this subject has been adequately covered by several authors who are possibly better qualified to do so than I am. The primary aim of this chapter is to acquaint the reader with information which will enable him/her to get the most out of his/her design. The information in question is concerned with the use of power MOSFETs as the power control elements in linear regulators, and as power switches in switch-mode supplies.

The power IGT will mostly be neglected in this chapter, since, in my opinion, it is at present not the device to be designed into modern high performance supplies whose switching frequency extends to several hundred kilohertz. The reader should be aware that the use of switching aid MOSFETs extend the use of IGTs into this area and information related to IGT switching aids will be covered later in the chapter. The MOSFET, on the other hand, requires few switching aids, if any, and is usually more cost effective by itself.

The derivation of transfer functions and the achievement of the necessary stability criteria should be of little concern to the reader, who is perhaps less interested in the academic side of the design phase and is looking towards gleaning some practical tips to ease the difficulties related to the development of switching regulators. If the reader should require documentation relating to the subject of stability, etc., then it is suggested that one of the many available textbooks should be consulted.

116   POWER SUPPLIES

Continuing with the theme of introduction it is also incumbent upon me to indicate where MGTs may be used in power supplies. This need arises out of the fact that the energy crises of the past two decades has forced efficiency to be placed amongst the highest of priorities. Bearing this in mind we can see that MHTs can be considered for application in three distinct areas of PSU usage: (a) the linear power control element in linear regulators, (b) the power switch in SMPS and (c) as low voltage high efficiency rectifiers in both linear and switch-mode supplies.

8.1 LINEAR REGULATORS

When the subject of linear regulators is raised the type of supply which is usually envisaged is the *series regulator*. With this type of regulator the control transistor is connected in series with the load. Those readers who can remember circuits containing vacuum tubes will be aware of a second type of regulator — this particular type being the *shunt regulator*.

The demise of the vacuum tube also signalled the passing of the shunt regulator, except for a few specialised applications. One of these special applications being the use of low power Zener diodes as voltage reference sources.

The reason for the lack of popularity of the shunt regulator, in spite of some attractive qualities, is twofold. The reasons for the virtual demise of the shunt regulator is predominantly due to limitations of the BJT. The limited stability of this device, at extended temperatures, along with the device's propensity to venture, at inopportune moments, into secondary breakdown meant that reliability of the supply would always be seriously questioned.

It has been unfairly alleged that low efficiency was one of the reasons for the lack of popularity for this type of regulator. Poor efficiency of the shunt regulator was never really the reason, since it can be demonstrated that a well designed shunt regulator can be virtually as efficient as a series regulated supply. In some respects the shunt regulator can be regarded as being more desirable than the series regulator.

The desirable aspects of the shunt regulator can be summarised as being:

(1) Inherent current limiting — without such a feature having to be specifically catered for either within the specification or having to be included in the design.
(2) Ripple reduction that is fundamentally superior to that provided by the series regulator. This feature also extends to the suppression of input to output noise transfer.

It is pertinent to now study the relevant advantages for including the power MOSFET as the power transducer in linear regulators when the strong point of the MOSFET has been its high switching speed. This factor (namely extreme rapidity of switching) is most certainly of little importance in this instance.

The MOSFET's strength for this application is its cost-effective use of silicon since the vast majority of linear supplies use power transistors rated at 60 V or less. At this breakdown voltage level the MOSFET, because of the higher current density at which it operates, requires less silicon for the same current-handling capability than the BJT. The efficiency of silicon usage has become even more important owing to the price erosion of the MOSFET, as evidenced in recent times.

Apart from the superior silicon utilisation and the unquestionably superior ruggedness of the MOSFET in terms of current handling ability and avalanche, the advantage of the power MOSFET over the BJT extends to three further parameters:

(1) Lower saturation on state voltage for equivalent dice area — thus lending itself to superior output/input voltage ratio and hence superior drop-out performance.
(2) Lower drive power requirements.
(3) Superior equivalent gain–bandwidth or transition frequency (f_t) enabling superior performance so far as transient response is concerned.

If the MOSFET happens to be one of the current sensing variety then another feature becomes available to the circuit designer. This is the ability to use low dissipation components for current sensing in the current-limit circuit.

Beyond the points raised relating to the benefits which may be accrued from sense or standard MOSFET usage, it should be pointed out that for normal linear regulators use of either bipolar or MOSFET will finally depend upon personal preference on the part of the designer.

The purpose of this book is not to enable the reader to design good power supplies to meet the requirements of mediocre specifications. It is primarily aimed at giving the reader some insights which will enable the design high performance supplies for lower overall cost.

8.2 LOW DROP-OUT HIGH EFFICIENCY LINEAR REGULATORS

The efficiency of linear regulators is related to the voltage drop across the series pass transistors. The voltage drop across this transducer is in turn a function of the dynamic range (as given in the specification) of the total input voltage excursion. If this dynamic range is too large and adequate head-room has to be allowed for, in order that regulation is maintained at $V_{input[low]}$, then this performance is achieved at the expense of poor overall efficiency.

118 POWER SUPPLIES

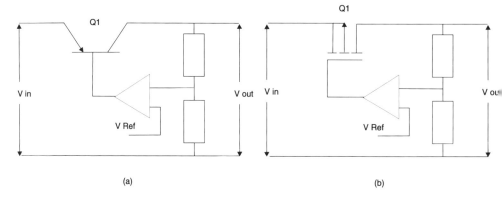

Figure 8.1 Linear regulators with low drop-out capability

One of the techniques which has been adopted as a means of overcoming this undesirable aspect of the linear regulator has been the use of pre-regulators and both linear and switching pre-regulators have been included. Overall efficiency is improved but this is achieved at the expense of cost, complexity and possible noise if switching technology is used for the pre-regulator.

Another alternative, and one which proves to be an acceptable compromise, is the use of low drop-out regulators. The term 'low drop-out', in this instance, signifies that the regulator is capable of regulating with a low voltage across the series pass transistor. Two types are illustrated in Figure 8.1.

The two circuits of Figure 8.1 both reveal the use of PNP/P-channel devices for the economic realisation of low drop-out regulators. The use of NPN/N-channel series elements has been deliberately resisted since this would have required the use of additional bias supplies, incurring both cost penalties and increased complexity.

In order to fully appreciate the benefit of MOSFET usage the reader should understand how drop-out voltage is defined. This may be given as:

The drop-out voltage of a linear regulator is that voltage difference between input and output below which output regulation at full load can no longer be maintained.

This can be expressed mathematically as:

$$V_{LDO} = V_{in[min]} - V_{out[max]} \qquad (8.1)$$

where $V_{in[min]}$ is the absolute minimum d.c. voltage which will be supplied to the regulator at full load, when full output regulation would still be a requirement. $V_{out[max]}$ will also be found within the supplies specification. If $V_{in[min]}$ is given as a d.c. value then this parameter should be inserted into equation (8.1). If, however,

8.2 LOW DROP-OUT HIGH EFFICIENCY LINEAR REGULATORS

$V_{in[min]}$ is given as an a.c. value (post-transformation) then the transformer secondary resistance needs to be specified. From all of this information the designer can discover the correct value for $V_{in[min]}$ from:

$$V_{in[min]d.c.} = (V_{in[min]a.c.}\sqrt{2}) - [V_{f(rect)} + (i_{L(max)}R_{sec})] \quad (8.2)$$

where $V_{f(rect)}$ is the maximum forward voltage of the rectifier at the nominal value of the maximum load current (twice this value should be used for full wave rectification by a bridge rectifier), R_{sec} is the secondary winding resistance of the transformer.

$$V_{in[min]} = V_{in[min]d.c.} - V_{ripple(pK-pK)} \quad (8.3)$$

where $V_{ripple(pK-pK)}$ can be found empirically from 10 000 μF per A (of load current) being required for the filter capacitor for 1 V pK–pK of ripple.

From equation (8.3) it is obvious that if the achievable value of L_{LDO} were maintained as low as possible then the lower the value of $V_{in[min]}$ would have to be to maintain regulation, and this would in turn lead to an overall improvement in efficiency. V_{LDO} for a PNP will tend to be greater than 1 V regardless of the size of the junction. For a P-channel MOSFET the value of V_{LDO} will be equal to the product of $i_{L(max)}$ and the maximum value of $R_{ds[on]}$ of the selected MOSFET. The design should allow either type of series-pass transistor to transit from its linear region of operation and to enter saturation as V_{LDO} is encountered. This can best be illustrated by the following example.

What would be the minimum input a.c. supply voltage required for a d.c. output of 24 V at a maximum output current of 2 A? What value of filter capacitor would be required for this application? Provide answers for both a PNP BJT and for a p-channel MOSFET. In both cases assume a centre-tapped secondary with single rectifier per winding limb. Also assume for both examples that $V_{f(rect)}$ is 0.7 V.

For a PNP (say the complement of 2N3055):

Assume $V_{ce(sat)}$ for the PNP at $i_{L(max)}$ is 1.0 V. Then V_{LDO} would be 1 V and $V_{in[min]d.c.}$ would be 25 V d.c.
Assume that the maximum pK–pK ripple voltage is 1 V then empirically C_{res} would be 10 000 × 2 μF, namely, 20 000 μF (say **22 000 μF**). $V_{in[min]}$ would be 26 V and $V_{in[min]a.c.}$ would be 26.7/1.414 = 18.88 V r.m.s. and after rounding up to a convenient value would be **19.0 V r.m.s.**

For a P-channel MOSFET (say IRF9Z34):

$R_{ds[on]max}$ (for worst case) would be 0.238 Ω; therefore V_{LDO} is 2 × 0.238 = 0.576 V. $V_{in[min]d.c.}$ would be 24.576 V and $V_{in[min]}$ would be 24.576 + 1 + 0.7 = 26.276 V. $V_{in[min]a.c.}$ would be 26.276/1.414 = 19.29 V, or **18.6 V r.m.s.** (after rounding). C_{res} would again be **22 000 μF**.

It is accepted that the examples given above are extremely simple, but there is unlikely to be an engineer who could not have worked out the answers without requiring the use of referral notes or a textbook for reference.

The improvement in efficiency n which the P-channel would offer over the PNP can be demonstrated to be:

$$n = 100 \times [V_{LDO(P\text{-channel})} - V_{LDO(PNP)}]i_{L(max)}/P_o \qquad (8.4)$$

where P_o is the output power.

Once again using the examples for the PNP and P-channel the improvement in efficiency from equation (8.2) will be 1.77% which should not be neglected as being insignificant.

The tendency towards higher efficiency inevitably means that the linear regulator is repeatedly being relegated to more specialised usage, and with greater emphasis being placed upon the switching regulator.

The reader should be aware that measurement of the linear regulator's efficiency is straightforward. One can use multimetres to measure input voltage and current. The product of these measurements will yield the input power. Similar measurements at the output will yield the output power. The linear regulator's efficiency will now be found to be the ratio of output power to input power.

If similar measurement techniques are applied to the measurement of input power for an off-line switcher and if the reader now calculates the efficiency surprising results may be obtained. The efficiency thus calculated may on occasion be found to be worse than that expected of a linear regulator when this clearly should not be the case.

The reason for the discrepancy is easy to ascertain. In the case of the off-line switcher, post-rectification filtering usually involves a large electrolytic capacitor. This type of filter results in the input line current having a non-sinusoidal waveform resulting in a severe peak to average current ratio or non-unity *form factor*.

I use the term 'form factor' to satisfy purists, because it is more usual, within some parts of the power supply fraternity, to refer to the switcher as having a poor *power factor*. The usage of the term 'power factor' to refer to the input of the switcher is slightly misleading since there is seldom no phase deviation between input voltage and current.

The reason that this form factor argument is customarily absent in the case of the linear regulator is the 50/60 Hz line transformer. Its considerable leakage inductance usually provides sufficient impedance filtering, to the input current waveform, to overcome the self-same electrolytic post-rectification filter. Overcoming the problem in the switcher involves some form of power factor correction either by active or passive circuit networks.

8.3 SWITCHING REGULATORS

It is not my intention to delve too deeply into the various topologies and their respective merits or otherwise. I would again say that this aspect of power supply technology has been covered by many other writers. (If the reader is not conversant with the various topologies then I would strongly recommend that he acquire any one of the books which have been published by Hnatek, Middlebrook and Cuk, Wood and Severns and Bloom amongst others. I am unble to single out any one of these authors.)

The topics which I will endeavour to cover will be concerned with such matters as frequency, 'snubbing', the topologies better suited to MOSFET usage, and whether resonance will achieve dominance over 'quasi square-wave' converters in the future. I will also present a few novel converter circuits and explain some of their features.

The 'switcher' will also tend to be the dominant theme of this chapter for the simple reason that it is less well understood and more frequently misjudged, when comparisons are drawn with linear regulators. It is also the largest single type of regulator that is produced by the 'merchant vendor'.

8.4 CHOICE OF FREQUENCY

I am frequently asked which range of frequencies will predominate in the design of the switching supply in the near future. This question arises because of the 'specmanship' which has to an extent become endemic among the vendors of power supplies over such matters as power density and therefore the volume of their respective products.

The higher the switch frequency the smaller will be the potential size of the magnetic components. This is true for *quasi-square-wave* converters. It is less true, for ultra-high frequency resonant converters where cores are operated at ridiculously low flux densities, that a quasi-square-wave converter can frequently achieve the same power throughput at one tenth of the switch frequency. It is relevant to mention this implication because of the potential connotations regarding the size of magnetic components being one of the factors affecting and determining the size of supplies. It is important that the reader be aware of this, since other factors besides frequency must also be taken into consideration when considering the overall design.

In off-line applications, the various international safety standards governing transformers relating to isolation voltage and also to creepage distances and clearances between the primary and secondary windings, and also to the core if it just happens to be grounded will ultimately create finite dimensions below which it will not be possible to proceed. It will be these finite dimensions which will ultimately limit the size of the magnetic components. Increasing the frequency beyond the value required to realise the size of the magnetic component will become merely academic and ultimately futile. The futility of working at frequencies above the limit set by safety legislation arises out of the increased losses of power switch(es), drivers and also within the magnetic components themselves.

Certainly the demands of aerospace and military contracts will appreciably reduce the volumes of power supplies below those which will be achievable by commercial off-line units. It is this market which will ultimately determine to some extent the favoured topologies and characteristics of switching regulators.

In the case of the commercial supply, frequencies of a few hundred kilohertz are already being adopted and this limit could well increase marginally for the off-line supplies. But the frequencies which have been and will be achieved are purely of academic interest to the engineer who has 'grown up' with the the BJT. This device is limited to the upper tens of kilohertz for the really fast devices, and then they are not at all competitive with the MOSFET on the basis of overall cost. The same frequency limits apply to those first generation IGTs and also to some second generation units because of that device's turn-off current tail. The latest second generation IGTs can be operated at these exotic frequencies if switching aid MOSFETs are used to reduce the switching losses. This allows the use of a combination of modest sized IGT and MOSFET against the requirement of a MOSFET of chip area which would equate to more than the combined silicon of the combination.

8.5 TOPOLOGY PREFERENCES AND THEIR IMPACT ON POWER SWITCH TECHNOLOGY

This section will not discuss converter topologies chronologically or their respective pros and cons. It is intended to reflect upon their effect on BJT development with respect to off-line SMPS usage primarily and to a lesser extent also to automobile ignition.

The inductive discharge ignition circuit may be considered as being essentially a Buck-Boost converter without secondary rectification, while the capacitor discharge circuit can be regarded as being a Buck derived circuit (again without secondary rectification).

Some of the more familiar converter circuits are illustrated in Figure 8.2. I have only included the most common circuits.

8.6 SINGLE SWITCH FORWARD/FLYBACK CONVERTER

One example of single switch forward/flyback converter circuit is shown in Figure 8.2(a) and has the following characteristics related to switch requirements.

8.6 SINGLE SWITCH FORWARD/FLYBACK CONVERTER

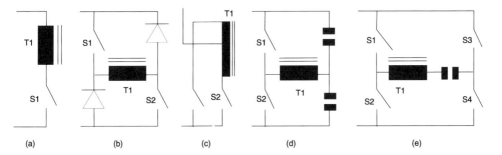

Figure 8.2 Basic converter topologies

Switch breakdown voltage is equal to twice the raw d.c. supply voltage if 50% conduction duty cycle is employed. This requirement arises out of the need to balance the volt-second products for both conditions of the power switch, namely ON and OFF. This twice supply voltage requirement resulted in BJTs of two dissimilar off-state voltage ratings — $BV_{ceo(sus)}$ and BV_{cex}. The latter voltage rating replaced in many instances the more usual one of BV_{cbo}.

Most of the transistors which 'sported' the BV_{cex} parameter were also found to have another 'desirable trait', namely the non-rectangular Reverse-Bias Safe Operating Area (RBSOA) as shown in Figure 8.3. This highly asymmetric curve for the RBSOA had a most profound effect upon the evolution of circuit concept development — the turn-off snubber. This sub-circuit was intended to keep the turn-off trajectory of the switch well within the RBSOA's boundary, by slowing down the voltag rise-time (t_{rv}) at the collector. This effect by the snubber upon t_{rv} has given an alternative name for the circuit within some circles of slow-rise network. One side-effect of the snubber is its tendency to reduce switching losses and thereby overall power dissipation within the switch. This side-effect has frequently resulted in the sub-circuit being utilised incorrectly and unfortunately for the wrong reasons.

The twice supply voltage rating for the power switch is not mandatory, but alleviation of the requirement means the abandonment of the 50% duty cycle operating condition, with all of the associated ramifications. The 50% duty cycle requirement disadvantages the MOSFET owing to the handicap of the $R_{ds[on]}$ being a function of $BV_{dss}^{2.5}$. This inefficiency can easily be overcome cost-effectively and methods to do so will be discussed later in this chapter.

The 750 V BV_{cex} rating requirement and the subsequent attempts at finding alternatives, to the single transistor converter, resulted in the development of the double-ended (two-transistor) forward/flyback converter circuit of Figure 8.2(b), which is sometimes erroneously called an asymmetric half-bridge. The circuit does overcome the 750 V requirement at the expense of using two transistors, two commutation diodes, along with additional drive complexity for the added dubious benefit of having to limit the duty cycle to less than 50%.

124   POWER SUPPLIES

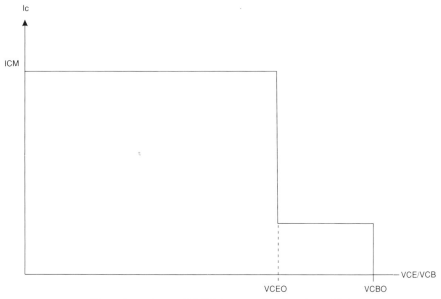

Figure 8.3 Typical RBSOA curve of high voltage BJT

The quest of push–pull operation possibly arose out of the misconception that transformer core utilisation would be significantly improved, compared with usage in the forward and/or flyback converter. This fallacious thinking arose out of the idea that the full B–H loop capability could be utilised with the push–pull, as against working around a minor loop (returning to $B_{remanance}$), in the case of the forward/flyback circuit. This is not entirely the case (although there is a certain element of truth in the assertion), since the core size will be determined by the losses, which in turn are found to be directly proportional to the operating flux density.

8.7 SYMMETRIC PUSH–PULL

The fundamental characteristics of symmetric push–pull topology is that the converter is Buck derived when operated in the voltage-fed mode and is boost derived when operated in the current-fed mode. The circuit is displayed in Figure 8.2(c). The voltage rating of the power switches must be equal to twice the voltage which appears at the centre-tap of the transformer. This requirement initially did not appear to be too much of a hindrance to the BJT, but was considered to be an embarrassment to the MOSFET as stated previously. The circuit did not gain significant success in acceptance; and this was certainly not a reflection on the voltage rating of the switch. The subtle 'hidden nuances' of *flux staircasing* and the subsequent need for *asymmetry correction* proved to be the biggest stumbling-block. The point relating to flux staircasing only applies to voltage-fed operation. In the current-fed mode this apparent disadvantage is found to be of no real consequence.

8.8 BRIDGE TOPOLOGIES

The drawbacks to the symmetric push–pull probably provided the spur to the development of bridge converter topologies. Certainly the drawbacks of the symmetric push–pull have resulted in the bridge circuits being very much more popular in high-power SMPS circuits. There is, however, an added proviso for this popularity. It is, that for all things being equal, the full-bridge which has the highest power throughput for a given transformer core. The two most common versions of bridge circuits are given in Figure 8.2(d) (half-bridge) and Figure 8.2(e) (full bridge).

8.9 600 V BV_{DSS} MOSFETs IN 220 V OFF-LINE FLYBACK SUPPLIES

It was stated previously that 750 V BJTs were not considered as being mandatory in this application provided that the conduction duty cycle was maintained at less than 50%. This statement, although being fundamentally correct, does not indicate that designs with BJTs with less than 750 V BV_{ces} can readily be achieved. Effects of leakage inductance and other strays usually require some form of voltage clamping to be utilised.

Transformers with *clamp/energy recovery windings* would perform almost all of the required clamping, but would do so at the expense of an increase in leakage inductance, and this would still require some other form of voltage clamp to be used.

Early MOSFET applications demonstrated the use of soft/spongy clamps as being eminently suitable, provided the extra cost did not prove to be unwarranted. Since this type of clamp is not entirely proof against over-voltage it also required the back-up of extraneous surge protection. The implication could well be that 750 V rated switches are an essential part for this topology. The solution is found to require a blend of circuit design and proven avalanche capability on the part of the MOSFET.

In order to fully explain the technique some knowledge of magnetic components is assumed on the part of the reader. This knowledge will enable a more meaningful understanding of the principles involved. The reader should remain assured that a relatively detailed explanation of the flyback circuits operation will be given.

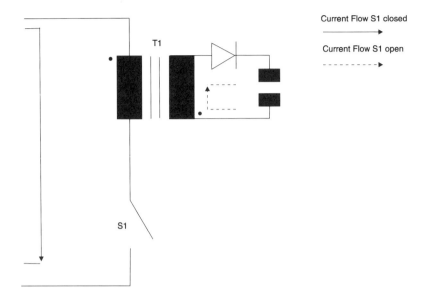

Figure 8.4 Simplified fly back converter showing primary and secondary currents and directions

In the circuit of Figure 8.4 the initial conditions which are deemed to be prevailing is that no current is flowing either in the primary or the secondary, and that switch S_1 is open. At some instant in time, given as t_o, S_1 closes and S_1 is assumed as being perfect with zero on-state resistance. Supply voltage V_{ds} is applied across L_{mag}, the magnetising inductance of the transformer T_1, and current commences to flow into the primary of T_1. (It is assumed that T_1's magnetic excursion commences at B = 0 and H = 0.) This primary current is the magnetising current for T1 and will commence to ramp-up in an almost linear fashion; owing to the effect of L_{mag}. No current will be found to be flowing in the secondary owing to the phase relationship of the secondary winding with respect to the primary and the connection and polarity of rectifier D_1.

At some time, determined by the current in T_1's primary and by the pulse-width modulator, S_1 opens. The magnetising inductance L_{mag} will endeavour to maintain the flow of magnetising current i_{mag} but because of S_1 being opened i_{mag} ceases to flow in the primary and flows instead in the secondary (and the voltage across the primary reverses), in an attempt to transfer the energy stored in L_{mag} to the output. This energy is given as:

$$\text{Energy in primary} = L_{mag} i_{mag}^2 / 0.5$$

All of the energy in L_{mag} is perceived as not being transferred to the output because C_1 is charged by the current in the secondary to a voltage which causes D_1 to be reverse biased. But this will not happen during the first switch cycle,

8.9 600 V BV_{DSS} MOSFETs IN 220 V OFF-LINE FLYBACK SUPPLIES

only during subsequent switch cycles. The energy which does get transferred can be equated as:

$$C1(V_{outs}^2 - V_{oute}^2) = L_{mag}(i_{magHI}^2 - i_{magLO}^2) \quad (8.5)$$

where V_{outs} is the start value of the output voltage V_{out}, V_{oute} is the end value, i_{magHI} is the maximum value of i_{mag} at the time S_1 opens, and i_{magLO} is the value of i_{mag} at the time D_1 becomes reverse biased.

The amount of energy which is transferred, although being of importance for the purposes of the design of the transformer, is of little relevance for the purpose of 600 V BV_{dss} MOSFET design-in. In this instance the value of V_{oute} is the all important parameter. The transformer should be designed such that:

$$N_{pri} \leqslant 600/V_{oute} N_{secy} \quad (8.6)$$

where N_{pri} and N_{secy} are the primary and secondary turns.

The ratio of N_{secy} and N_{pri} can be in turn given as N, so that:

$$N = V_{oute}/600 \quad (8.7)$$

We now have a system of auto-clamping where the output filter capacitor C1 clamps the drain voltage of S1, and provided the pulse-width modulator balances the volt-second products the circuit will be found to perform with perfect reliability. The only remaining problem is the energy stored in the leakage inductance L_l which will cause a transient spike to be superimposed upon the drain voltage waveform and equates to:

$$L_{l[energy]} = l_l i_{magLO}^2 / 2$$

Provided the value of $L_{l[energy]}$ is within the E_{ar}, the repetitive avalanche rating of S_1, and provided the product of $L_{l[energy]}$ and f_o, the switch frequency, does not cause the overall dissipation in S_1 to exceed the maximum t_j of the transistor it can be seen to have been demonstrated that the combination of an avalanche rugged MOSFET and snubberless operation are perfectly possible. If used sensibly this combination will be seen to be inordinately cost-effective.

The auto-clamping action is not necessarily confined only to the flyback converter. It can also be applied to other types of converter. One of these which is also quite novel in concept is given in the schematic of Figure 8.5.

The converter circuit of Figure 8.5 combines several unusual features in an extremely novel concept. The combined features include:

(a) No snubbing required for S_1 due to the auto-clamping action of T_1.
(b) The two transformers T_1 and T_2 could be integrated on to one core if so desired. This will be illustrated in Figure 8.6.

128   POWER SUPPLIES

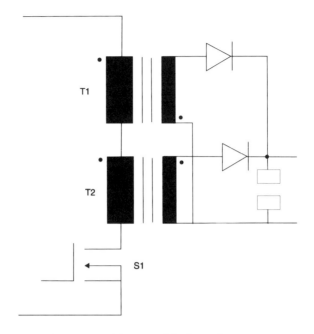

Figure 8.5 Schematic of fly-forward converter

(c) It is truly unique amongst converters in having the ability for a single switch topology to transfer energy from primary to secondary during both states of S1, i.e. when closed and open. Purists might take exception to this last statement for there is the single-switch series resonant supply which has found a certain degree of favour within modern microwave ovens. These supplies accomplish energy transfer during both states of the power switch. My original assertion still stands since the series resonant supply stores energy in the core of the transformer during switch closure and then subsequently releases this energy when the switch opens.

(d) Design simplicity.

The stages of its evolution can be considered as follows:

(1) Transfer the output filter inductor from the secondary side to the primary side. This first step may be considered as having created a converter akin to a flyback converter, but with the incorrect phase relationship between transformer windings. (The reader who may be a first-time designer of a SMPS, or someone who lacks a great deal of experience, may be perplexed at this last claim, since all schematics of a flyback converter will depict no inductor on the primary side. I would draw their attention to the real and well defined magnetising inductance within the transformer. The magnetising inductance needs to be much more clearly defined with this converter than for any other.)

8.9 600 V BV_{DSS} MOSFETs IN 220 V OFF-LINE FLYBACK SUPPLIES

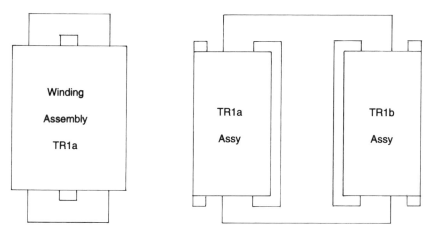

Note:
TR1a Assembly on core limb with air-gap

Figure 8.6 Magnetic core details of transformer for the fly-forward converter

(2) Change the construction of the inductor, now on the primary, to be a flyback transformer. The commutation (free-wheeling) diode of the forward converter becomes the rectifier of the flyback half.

The design simplicity which is so fundamental to this interesting circuit is enabled by the strict adherence to the few simple rules which make up the design process of the transformer.

The physical details relating to the windings and to the magnetic core are given in Figure 8.6. The rules concerned with determining the real values of primary and secondary turns may be stated as follows:

(1) Initially make $N_{p1}/N_{s1} = N_{p2}/N_{s2}$ and also equal to N (where N_{p1}, N_{s1}, N_{p2} and N_{s2} are the primary and secondary windings of transformers T_1, and T_2 respectively.)
(2) The auto-clamping action may be determined from:

$$V_{DS(off)} = e_i + e_o(N_{p1}/N_{s1} + N_{p2}/N_{s2}) \tag{8.8}$$

$$V_{x(off)} = e_i + e_o N_{p1}/N_{s1} \tag{8.9}$$

where $V_{DS(off)}$ is the clamped off-state voltage of S_1, $V_{x(off)}$ is the voltage measured between N_{p1} and N_{p2} for S_1 off and e_i and e_o are the input and output voltages (of the supply) respectively.
(3) For the application in question and using equation (8.8) calculate N for both transformers. For off-line applications and for 600 V MOSFETs make $V_{x(off)}$ equal to $e_i + 100$ V.
(4) Having determined N and knowing the value for i_o (the output current), the peak value of primary current can be calculated. This value is also the

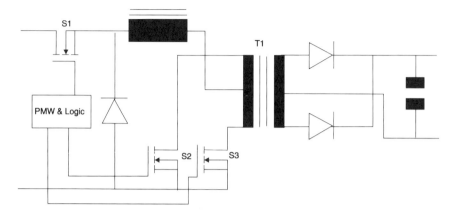

Figure 8.7 Schematic of the current-fed push–pull converter

magnetising current in T_2 and enables the magnetising inductance for this transformer to be calculated using the A_L factor for the selected core. With the magnetising inductance and therefore the primary turns for T_2 having been calculated, the secondary turns can be calculated.

(5) From

$$-e_i - V_{x(on)} = L_{mag1} di_{pri}/dt$$

Calculate the primary inductance of T_1 and again, using the A_L factor of the core, calculate the primary turns. (Ensure that the Li^2 capacity of the core has not been exceeded by the use of a suitable gap in the core if necessary.)

(6) Calculate the secondary turns of T_1. It may be necessary to repeat steps (1)–(6).

The symmetric push–pull converter of Figure 8.2(c) when cascaded with a Buck converter furnishes another interesting circuit, namely the current-fed converter of Figure 8.7. The topologies given so far are not the only ones with the intrinsic capability to promote the benefit of auto-clamping. Almost any type of converter, where the output voltage is not blocked from the secondary by finite inductance, can be configured to have this feature.

Besides having the ability to auto-clamp the drain voltages of the power switches the circuit has other in-built features which are important enough by themselves to warrant the inclusion of the converter for the purposes of discussion. The pre-regulatory nature of the circuit infers that the output of the secondary winding will be a very stable source of a.c. power. If more than just one secondary winding (and output) is therefore included, the cross-regulation between outputs is found to be perfectly acceptable.

The stable source of a.c. power means that small high-frequency transformers with simple rectifier and filter circuits may be deployed in a distributed manner throughout a major item of equipment and thereby ensure that localised rectification and filtering are all that is required within the various parts of the equipment.

8.10 EFFECTIVE TURNS RATIO VARIATION

In order that power supplies are able to operate from sources of widely varying voltages, it is sometimes desirable if the wide-band transformer could incorporate a measure of tap-changing as part of the overall circuit. Several converter circuits are currently in existence which utilise this or a similar technique.

The current-fed push–pull converter of the previous section can be included within this section if desired, although it is fair to delineate this converter from true tap-changing converters.

A third set of converter types utilise a novel technique to effectively vary the primary voltage which is applied to the primary of the transformer. Two of these converter circuits are included in this chapter, since in both instances the high peak to average current capability of MGTs tends to be a distinct advantage. It is fair to point out that both converter circuits have been originally developed to use BJTs.

The first of these converter types has the limitation of being able to vary the primary voltage by a mere ratio of 2:1. This variation may on occasion prove to be insufficient for the type of application for which it was intended. If a larger ratio is required it may be necessary to consider the other alternatives.

The converter circuit in Figure 8.8 has the capability of operating in one of three definite modes. The mode which should be considered first is similar to that of a conventional half-bridge. Only switches Q3 and Q4 are in use with switches Q1 and Q2 being more or less permanently inoperative in an open-circuit condition. Automatic clamping by capacitors C1 and C2 'à la half-bridge' ensures that the voltage applied across the two primary windings equate to half the raw supply voltage.

The maximum voltage therefore applied across the power switches in their off-state is half the supply value across Q1 and Q2 and equal to the full value of the supply across the switches Q3 and Q4.

In the second of three modes, switches Q1 and Q2 are switched diagonally in synchronism with Q3 and Q4 'à la full-bridge' and the circuit is deemed to be operating virtually as a full-bridge. The voltage applied at any instant in time across the transformer primary can be shown to be equal to the raw supply voltage. Since the source/emitter or drain/collector of the off switch is always clamped through its associated diode to the centre point of capacitors C1 and C2 it can be demonstrated that the maximum voltage which is applied across either Q3 or Q4 is 50% greater than the supply voltage in value.

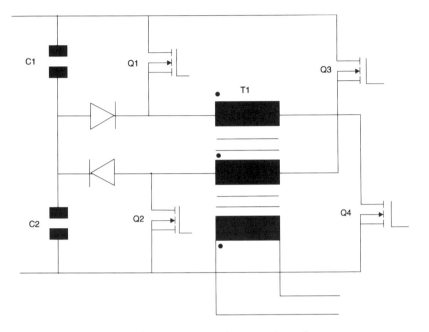

Figure 8.8 Novel converter circuit for step voltage changes

In the third of the operational modes Q1 and Q2 can be switched on permanently with only Q3 and Q4 being pulse-width modulated. Again the voltage which is applied across the transformer primary windings is found to equate with the supply. It can now be demonstrated that the voltage applied across either Q3 or Q4 (during the off-state) is equal to twice the supply voltage in value. In this third mode of operation the circuit may be considered as being similar to a pair of asymmetric half bridges operating in parallel.

The only requirement for the circuit to be able to change its operational mode is that the control circuit be able to perform the transition during the changeover from one primary winding being in circuit to the take-up by the other primary winding.

A variation in the operational parameters for this circuit would be to use bi-directional synchronous rectifiers on the secondary of this converter in order to generate a synthesised sine wave.

One of the advantages of utilising the approach advocated by the stepless converter is that regulation may be controlled by a microprocessor so that the frequency can be easily changed. It could conceivably be a few instructions in the software code.

The second of the two circuits having the ability to vary the voltage which may be applied across the primary winding of the supply's transformer is given

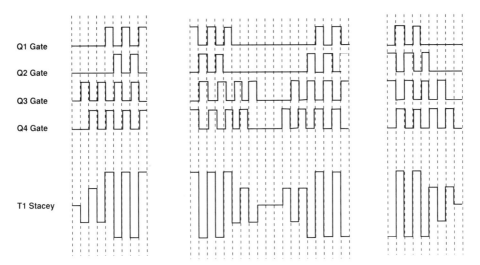

Figure 8.9 Waveforms of the converter of Figure 8.8 generating a synthesised sine wave output

in Figure 8.10. In this converter the circuit is almost that of the half-bridge with the only change being to the connection of the 'splitter capacitors' C1 and C2. Switches Q1 and Q2 operate at a fixed duty cycle of 50%. It is only Q3 which operates under pulse-width modulation control. Pulse-width modulation of Q3 may be viewed as being similar to varying the resistance of the connection between the splitter capacitors.

This converter, unlike that of Figure 8.8, has the ability to vary the voltage applied across the transformer primary over a wide range and in an almost infinite number of steps. It is also therefore quite likely that this converter is fundamentally superior to that of Figure 8.8. The two circuits have been illustrated so that the user may decide in favour of whichever one is best suited for the particular application being worked on.

8.11 MGTs AS RECTIFIERS IN POWER SUPPLIES

The final section in this chapter is the study of MOSFETs as high efficiency synchronous rectifiers in low voltage power supplies.

The demand for MOSFET rectifiers is evident when the efficiency of supplies for TTL applications are studied — 5 V outputs. The situation deteriorates rapidly as supply output voltages go down towards the levels used by emitter coupled logic (ECL) as used in large mainframe computers. ECL supplies are down to 2 V and several hundreds of amperes.

A casual glance at rectification efficiency calculations will bring the problem into focus. Rectification efficiency may be defined as the ratio between rectifier forward voltage and the output voltage of the supply. This can be expressed thus:

134   MOTOR DRIVES AND CONTROLLERS

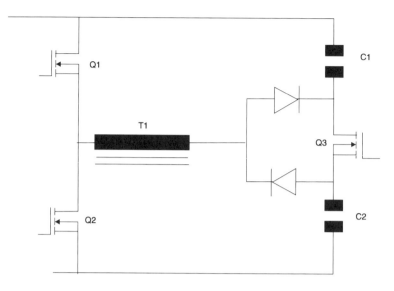

Figure 8.10 Half-bridge SMPs with splitter resistance control

$$n_{\text{RECTIFIER}} = 100[(1 - (V_F/V_{\text{OUT[PSU]}})] \tag{8.10}$$

Equation (8.10) applies with equal validity whether the rectifier diode is used as a rectifying element or as the free-wheeling diode.

We can now carry out a simple everyday exercise. Let us examine the implications of equation (8.10) in a power supply having a 5 V, 20 A output and for simplicity make the converter a forward converter which uses both a rectifier and a free-wheeling commutation diode. Let us also assume that the two diodes are standard Schottky rectifiers with a forward voltage (V_F) of 0.5 V, 20 A.

Using equation (8.10) we find that $n_{\text{RECTIFIER}}$ is 90%. If we assume that the rest of the supply is perfection personified in having zero losses we conclude that the supplies efficiency can never be better than 90%. Parallelling two rectifiers for each element will only yield a small improvement.

Let us now replace the individual rectifiers each with two MOSFETs of type IRFP054 connected in parallel. The 25 °C $R_{\text{ds[on]}}$ is 0.0075 Ω (max) with the 100 °C $R_{\text{ds[on]}}$ being 0.015 Ω. At 20 A, V_F will be 0.3 V. Once again, using equation (8.10) yields an efficiency of 94%. The conclusion must be that standard Schottky rectifiers place a limit upon the efficiency which may be achieved in any supply having a 5 V output.

In order to use MOSFETs in this application it is necessary to examine the best way in which they may be used. Examination of their characteristics indicates that it is not merely a question of using the parasitic as the sole rectifying element. The channel must also be brought into play. It is in this context that it becomes obvious that the MOSFET must be used as a switching type of rectifier, where high frequency is not the qualification.

Synchronous rectification with MOSFETs

In the quest for ever higher efficiencies one of the solutions which is under appraisal is the use of MOSFETs as synchronous rectifiers. The meaning here is that the channels enhancement is synchronous with the applied voltage. A circuit which demonstrates the use of MOSFET synchronous rectifiers is given in Figure 8.11.

Since MOSFETs are inherently resistances when fully enhanced into their on-state, the connection of several devices in parallel will enable the fabrication of rectifiers with extremely low forward voltages. The IGT, and for that matter the BJT, does not share this capability, because of the intrinsic voltage off-set in the on state.

It is equally true that ultra low-voltage Schottky rectifiers are available which enable this particular type of rectifier to raise a serious challenge to the MOSFET on the grounds of economic viability. It is only when the ultimate efficiency is sought, regardless of cost, that the MOSFET synchronous rectifier emerges as the clear winner.

Resonant or quasi square-wave converters

With the apparently continued pressure of increased switch frequency the question of choice of waveform becomes ever more pertinent.

The finite recovery time of rectifiers, and therefore their subsequent recovery losses, may prove to be unacceptable in terms of thermal management and efficiency. This also applies to a lesser extent to the power switch(es). Both BJTs and IGTs are found to operate quite satisfactorily at frequencies of several hundred kilohertz provided they have been installed in resonant or quasi-resonant types of converter. MOSFETs on the other hand are not in any way limited by such a constraint and can function with equal speed up to a few megahertz in a square-wave converter.

It has been demonstrated that the BJT is also capable of operating efficiently at several hundred kilohertz in square-wave converters provided the precaution of emitter switching has been included.

The reasons for the reader's interest in the type of converter must be not in the capability of the switch but in the type of converter that is best suited for highest efficiency at very high switch frequencies. It is not possible to pinpoint with any certainty which mode of operation is best suited to this particular aspect. Certainly, square-wave converters can be built to operate very efficiently at extremely high frequencies. I have seen demonstrated a forward converter deliver an output power of 1 kW at a frequency of 1 MHz whilst having an overall efficiency greater than 85%. Unfortunately the degree of expertise required

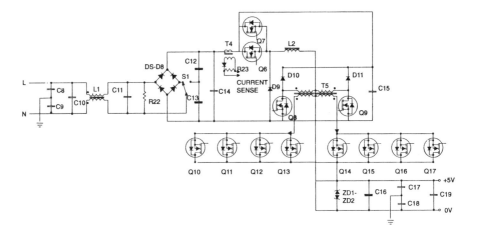

Figure 8.11 Power supply with MOSFET synchronous rectifiers (Courtesy of International Rectifier Corporation)

to achieve this possibility will be well beyond the capabilities of almost all first-time users and also beyond the capabilities of several seasoned users of MOSFETs.

It is probably in the best interests of all concerned if some form of table were to be drawn up listing the majority of the pros and cons of the two modes of operation. Table 8.1 does not list all of the major attributes because of limitations of space.

Table 8.1 Advantages and disadvantages of converter modes.

Resonant	Square Wave
1. More manageable at very high switch frequencies.	1. Less manageable at very high switch frequencies.
2. Pulse-width modulation not easily accommodated.	2. Pulse-width modulation is easy to accommodate.
3. Lack of (2) above implies frequency modulation which increases EMI/RFI filtering difficulties above earlier expectations.	3. Fast transitions can make EMI/RFI suppression more difficult to achieve. Increases susceptibility of switch oscillations.
4. High efficiency rectifiers easily acquired.	4. Rectification efficiency not easily achievable.
5. Control circuit difficult to configure.	5. Control circuit fairly easy to configure.
6. Does not accept large load variations easily. Requires some extra circuitry.	6. Can be made to easily cope with wide load load variations.
7. Magnetic core utilisation is not always optimised	7. Magnetic core utilisation is not a criticism.

8.11 MGTs AS RECTIFIERS IN POWER SUPPLIES

Since the details in Table 8.1 tend to be subjective, valid objections could be raised over the omission of some points. The statements made concerning potentially poor core utilisation within resonant converters is well known but frequently misunderstood and will not be one of the subjects of discussion since this virtually requires a course of instruction in the science of magnetic materials.

9
Motor Drives and Controllers

It is not my intention to delve into the constructional details and the niceties, or otherwise, of the various types of motors, for any one of which the design of a controller may be required. Any reference to the types of motors are included to demonstrate the type of power switch and the topology of the drive which may be required.

The *raison d'être* of this chapter is to indicate the various drive requirements of certain types of motor and any other problem areas which might be encountered.

The different types of motor, which could be considered as being suitable for control by means of modern electronic controllers, may be categorised as follows:

(1) Brushed d.c. motors — with single, dual or four quadrant control.
(2) Brushless d.c. motors.
(3) Multi-phase induction motors.
(4) Reluctance motors.
(5) Universal motors.

MGTs do not encounter serious difficulties when they are installed within motor controllers *per se*. The IGT without an integral *anti-parallel diode* has the inconvenience of requiring a discrete external diode to be connected across it. This is certainly true when the power stage of the circuit is made up of one or more totem-pole connected switches. To offset this inconvenience a choice of diode is available to the designer. The MOSFET does have the benefit of having an in-built commutation diode and with low voltage parts few if any problems should be encountered.

Unfortunately, MOSFETs with breakdown voltages of over 500 V are restricted to having a parasitic diode with a limited t_{rr} (*reverse recovery time*) and also from an ever increasing value of $R_{ds(on)}$. Devices having diodes with modest t_{rr} ratings, and which also lack ruggedness, will require some precautionary measures to be taken during the design phase if failure of the parasitic diode is to be avoided.

140 MOTOR DRIVES AND CONTROLLERS

One of the causes of parasitic failure is the onset of emitter injection occurring as a result of the recovering parasitic diode being exposed to a high dv/dt being applied across the device during t_{rr}. Once emitter injection has been initiated the possibility of second-breakdown occurring within the parasitic diode becomes a real threat. Another cause of failure in the parasitic diode is excessive dissipation arising out of the recovery phenomenon in rugged MOSFETs. The designer can reduce the possibility of failure owing to the second of the alternatives by selecting the device with the lowest *reverse recovery charge* (Q_{RR}). It is this characteristic which determines recovery dissipation in rugged MOSFETs which operate at a given frequency.

Alternatively MOSFETs with fast recovery parasitic diodes are available. These devices are frequently given the misnomer of *FREDFET*, where the FRED part of the name is the well known acronym for Fast Recovery Epitaxial Diode. The misnomer results from the fact that the parasitic NPN is alluded to as being something it is not. These devices are not the panacea which they initially appear to be. Their increased rapidity of reverse recovery is the result of *minority carrier life-time* control. The control of the life-time of the minority carrier by heavy metal doping has the tendency of making the t_{rr} speed of the device to be very temperature dependent.

Before the use of MOSFETS with fast parasitics are considered it would be advisable to study the types of circuits where they are likely to be needed.

Certain motors and the technique used to drive them require the use of a totem-pole transistor output stage with pulse-width modulation being used to control the waveform appearing at the centre point of the totem-pole. The favoured outline of the current waveform is usually regarded as having the desired shape of being sinusoidal. The requirement of a sinusoidal current waveform for single and multiphase induction motors is based on the premises of quietness of operation, smoothness over the complete r.p.m. range and the freedom from *cogging* at extremely low speeds. The term 'cogging' refers to the jerky motion of the motor at low speeds which may be observed when the motor is driven by a six-step inverter.

Modern MOSFETs without fast recovery parasitics do not suffer from dv/dt failure arising out of second breakdown in the parasitic. Their failure results from the losses incurred during recovery. In six-step inverters the total losses of the parasitic due to t_{rr} tend to be small, since the frequency of switching tends to be low. It is in PWM inverters that the parasitic's losses become meaningful. The energy per pulse loss in the parasitic is a function of the peak reverse recovery current i_{RM} and is the same for a six-step or PWM inverter. The power loss due to t_{rr} is expressed as the product of the energy and the frequency is given by:

$$\text{Power } (T_{rr}) = \text{energy per pulse} \times \text{switch frequency} \quad (9.1)$$

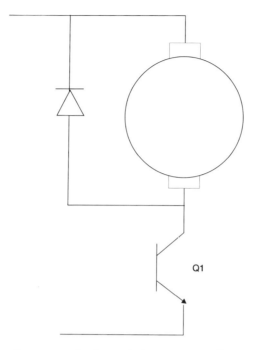

Figure 9.1 Single switch d.c. motor controller

It can therefore be seen that the ultrasonic PWM inverter, which utilises two or more totem-pole power stages, will place a greater burden upon the parasitic diodes within the MOSFET's structure than conventional six-step inverters.

The inference from the above paragraphs is that the end user of the motor must accept the inconvenience. Nothing could be further from the truth and to demonstrate this requires a certain amount of discussion concerned with the individual types of motor.

Consider the requirements for a brushed d.c. motor. First, a simple on/off controller may be all that is required for single quadrant control. Braking of the motor, if required, may be reduced to the simplicity of a mechanical brake.

A circuit which fulfils this requirement is illustrated in Figure 9.1. The gate drive for Q1 could easily be as simple as is needed to turn the MGT on and off, if this is all that is required of the motor. If speed control is the requirement then the same basic circuit will be found to be perfectly adequate with the condition that a PWM signal has to be applied to the gate of Q1. It is also a requirement that a commutating diode be connected across the motor in order to *free-wheel* the motor current when Q1 is off. The recovery characteristics of the commutation diode are determined by the frequency at which Q1 is switched. Complications arise when two-quadrant speed control is needed to be fitted to the brushed d.c. motor.

142   MOTOR DRIVES AND CONTROLLERS

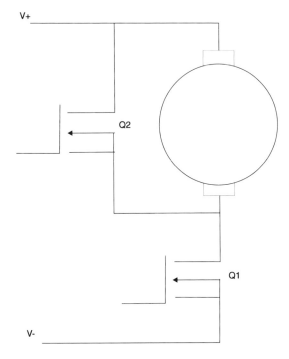

Figure 9.2 Quadrant control

Regenerative or dynamic braking of the motor calls for the subtle changes which are made to the controller which is shown in Figure 9.2. The commutation diode for the motor in Figure 9.1 has been replaced by the second MGT (Q2). Control of braking is achieved by PWM signals being applied to the gate of Q2. One of the major benefits which is associated with the circuit of Figure 9.2 is the ease with which a change in rotation can be achieved, always assuming that the motor has been series connected. Moving the second power connection of the motor from V+ to V− is all that is required, and not a complete reversal of the power connections. It is also true that an exchange of the logic functions of Q1 and Q2 will also be required.

The advantage of the totem-pole connection has been exploited in multiphase induction motors and brushless d.c. motors by using at least three totem-pole power stages, for a three-phase motor, as illustrated in Figure 9.3. A non-PWM step inverter fabricated out of MOSFETs with rugged parasitic diodes, i.e. diodes having some avalanche capability, will have few problems. MOSFETs with diodes having no avalanche capability might fail for no apparent reason.

The drawback of the circuit arrangement is the reverse recovery characteristics of the parasitic diode of Q2 in Figure 9.3 (if it is a MOSFET). At frequencies above 20 kHz rugged MOSFETs run into problems associated with the heat

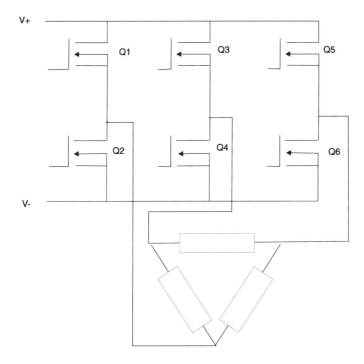

Figure 9.3 Three-phase bridge motor drive using power MOSFET

generated by the parasitic. In the case of MOSFETs without significant avalanche capability failure would occur even at low junction temperatures. The failure in this instance would almost certainly be associated with dv/dt *induced emitter injection* of the parasitic ultimately leading to *second-breakdown* within the parasitic itself. Several solutions for overcoming the problem have been suggested.

I touched briefly upon some of the subject material at the start of the chapter. I now feel that it is prudent to expand the initial material.

The first of these solutions was to modify the fabrication process for the MOSFET itself. This involved the use of heavy metals as dopants to reduce the life-time of the injected minority carriers. (The first-time user of MOSFETs should remember that it is only 'body diode' conduction that involves the flow of minority carriers.) These parts are usually sold under the title of FREDFET or HYPERFET. These devices alleviate the problem of the reverse recovery phenomenon but at the expense of increased losses owing to an increase in $R_{ds[on]}$. The introduction of life-time control has reduced t_{rr} to about 200 nS at 25 °C. Carrier life-time control has a secondary undesirable characteristic. This is the falling-off of reverse recovery speed with increasing temperature.

Because of the FREDFET's lack of total success in providing a definitive solution other techniques have been adopted. The more successful approaches

144 MOTOR DRIVES AND CONTROLLERS

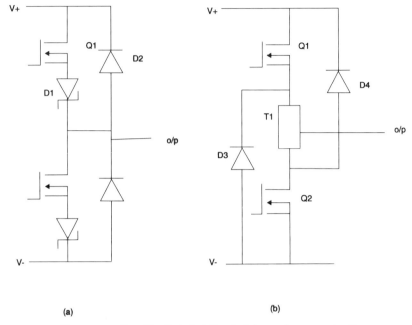

Figure 9.4 Parasitic diode isolation in totem-pole power circuits

have centred around total or partial isolation of the parasitic diode. Other techniques have attempted the reduction of applied dv/dt to the MOSFET during recovery of the parasitic, by the use of additional signal conditioning circuitry.

The circuits of Figure 9.4 indicate two methods at full or partial isolation of the parasitic. It should be stressed that these are but two of the methods which may be used.

An alternative approach is to actively limit the dv/dt which is applied to the parasitic which is recovering. The various methods are all based upon slowing down the turn-on of the opposing switch in the totem-pole arm of the PWM inverter.

The circuit of Figure 9.4(a) depicts the total isolation of the parasitic by including a Schottky diode D1 in series with the source of the MOSFET and a separate anti-parallel commutation diode D2. The circuit of Figure 9.4(b) approaches the problem from a different angle. In this instance the centre-tapped *auto-transformer* limits the peak reverse recovery current (I_{RM}) through the parasitic along with limiting the applied dv/dt. Anti-parallel diodes D3 and D4 are required for the transition of Q1 being on and turning off. D3 must now commutate the load current until T1 has moved from $+\beta_{sat}$ to $-\beta_{sat}$ and the parasitic of Q2 turns on. It should now be apparent that the average current through D3 in Figure 9.4(b) is significantly less than the average current through D2a in Figure 9.4(a). The selection of the core size of T1 is achieved by satisfying the equality of the equation where:

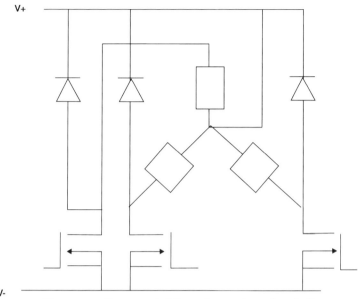

Figure 9.5 Star connected three-phase motor and controller

$$et = NA_E \delta\beta \qquad (9.2)$$

where e is the voltage applied across one half of the centre-tapped winding of T1 and is equal to V_{supply}, t is the reverse recovery time (t_{rr}) of the parasitic and may be found in the data sheet, N is the number of turns of one half of the centre-tapped winding, A_E is the effective area of the core and $\delta\beta$ is the peak-to-peak flux swing from $+\beta_{sat}$ to $-\beta_{sat}$ and may be found in the data for the selected core.

With experience the reader will become accustomed to selecting cores of relatively small mass and hence area. The calculations are not reiterative and are therefore not time-consuming.

Multiphase induction motor and brushless d.c. motor functions would benefit if an altogether different approach were adopted. In this instance it has nothing to do with the power stage of the electronic circuitry but is concerned with the motor itself. The two types of motor being discussed here are normally shown connected in a delta configuration. The connection is due to the internal connection of the motor. (I am aware that there are good reasons for the choice of winding connection.) My recommendation is valid if the type of connection used tends to be of little importance.

In this instance the only requirement for a cost-effective solution is for the motor to be connected in a 'Y' or star configuration. The complete solution in this case is given in the circuit of Figure 9.5. The reader should note that in this circuit the controller has been simplified considerably and has therefore become

146   MOTOR DRIVES AND CONTROLLERS

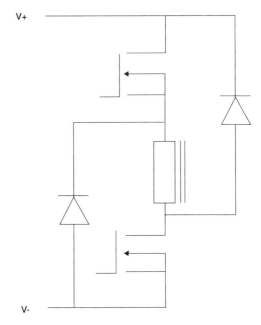

Figure 9.6 Drive for a single winding of the switched reluctance motor

very much more cost-effective, only requiring three MOSFETs instead of the more usual six items which are normally needed for a bridge inverter. It is accepted that three commutation diodes are an additional requirement but it should be remembered that these parts are less expensive than either type of MGT and an added benefit now is the lack of t_{rr} associated problems relating to the parasitic diodes of the standard MOSFET which has been covered elsewhere in this chapter.

A secondary statement relating to the 'Y' or star connection principle is the alteration to brushed d.c. motors. The concept referred to applies to those motors which are sometimes termed bifilar motors. It is not intended to cover the principle behind these types of motors but to make the reader aware of their existence. Bifilar motors require only two switches and two commutation diodes to achieve four-quadrant control.

The reluctance motor also requires a drive which is significantly different to either an induction motor or the brushless d.c. type. The controller and winding for a single winding reluctance motor is given in the circuit of Figure 9.6. This type of motor shares many of the features and is similar to the *stepper motor*. It is for this reason that I have not included the latter. The reader's attention is drawn to the similarity of the circuit to the two-transistor converter circuit which has found favour in the power supply market-place.

Universal motors are widely used within the consumer industry and may be found in an extremely diverse range of appliances. Some of these appliances

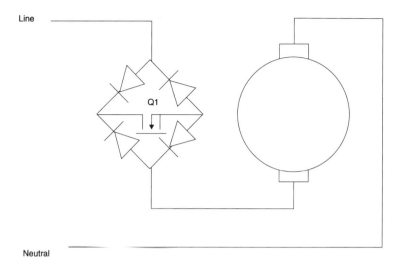

Figure 9.7 Universal motor speed controller

are: (a) washing machines; (b) hand tools; (c) vacuum cleaners and (d) dishwashers.

The four appliances which I have listed most certainly do not constitute the entire list but indicate the reasoning for introducing electronic control in conjunction with the universal motor. Certainly portable hand-drills have been available with electronic speed control. This relatively simple controller utilises a Triac operating in the *phase control* mode. Manufacturers of some of the other appliances which I have listed are endeavouring to apply sophisticated control to the speed of the universal motor.

Two of the appliances where sophisticated speed control is considered as being more than desirable are vacuum cleaners and washing machines. The requirements for the two appliances are completely different, although both functions may be regarded as being applicable to safety.

Vacuum cleaners with motors of 1 kW rating are becoming relatively commonplace. The marketing personnel associated with these appliances are now seeking to introduce models with 1.5 kW and even 2 kW motors. The reasoning behind this marketing decision is that the housewife will benefit with constant strength of suction, especially if the speed and torque of the motor are accurately controlled. The reasoning so far would appear to be totally valid. What the marketing personnel have failed to disclose is the safety hazard such appliances pose to the unsuspecting user, and to their clothing.

If the vacuum cleaner is returned to it soon becomes apparent that the simple phase control Triac solution is now found to have serious limitations.

A circuit that overcomes most of these limitations is given in Figure 9.7. A single MGT provides the most economical solution. The requirement for the

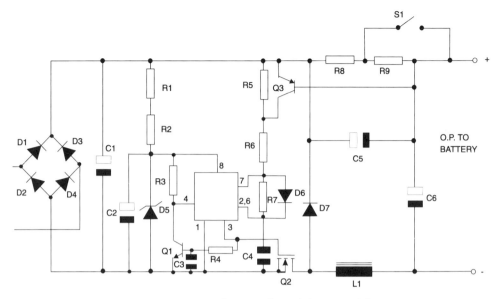

Figure 9.8 Battery charger with rapid charge capability

bridge rectifier is to overcome the unidirectional nature of present day MGTs. It would be perfectly feasible to replace Q1 with a BJT. The reason for the overwhelming superiority of the MGT is drive power. A BJT in the position of Q1 could conceivably require a line transformer to supply the power for the control circuitry. A simple *dropper resistor* and Zener is sufficient for the purposes of powering the MGT's control circuitry.

In the case of washing machines it is not merely the well-being of delicate fabrics which are being catered for by the prudent use of accurate speed control. Wash programs of ever increasing complexity and diversity are made possible by the advent of the microprocessor. Speed and directional control now become increasingly important. Marketing demands for ever-increasing spin speeds might herald the demise of the universal motor for this particular application. It is difficult to envisage an economical electronic controller which would provide all of the functions (pole reversal for directional change and tap changes for speed control) which are presently fulfilled by relays.

A motor and its controller which I have failed to cover adequately is the brushed d.c. motor which is the mainstay of the rechargeable battery shaver. The circuitry which has been used quite successfully in the past has been the capacitor attenuator and rectifier. This circuit has made rapid charging of the battery difficult to achieve in an efficient manner.

MOTOR DRIVES AND CONTROLLERS   149

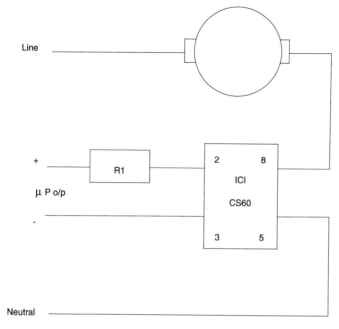

Figure 9.9 Shaded pole motor controller

The rapid change in the pace of everyday life coupled to the demands made upon the designer, by the supplier's marketing department, has led to the slow demise of the simple capacitor circuit. The battery charger has rapidly become a d.c.–d.c. converter. Constraints of cost has led to an intriguing circuit which is shown in Figure 9.8. Purists will recognise the converter as belonging to the Buck converter family and will note that the lack of transformer isolation may create protestations of rules of safety being breached. I would like to point out that most users of electric shavers use proprietary wall sockets which include transformer isolation. I would also like to point out that there are battery shavers in existence which make it a near impossibility to shave while the battery is being charged. The shaver has two retractable blade connectors which connect the shaver to a wall socket. I have such a shaver in my possession and I challenge any person to prove to me that it constitutes a hazard to my well being.

There are purists who will claim that the circuit of Figure 9.8 should have been included in Chapter 8. I have included it here as a motor speed controller because that is the role it could easily fulfil in the absence of the battery being omitted.

Another motor which I omitted from my original list is that miniature part that may be found in washing machines and dishwashers connected to the scavenge pump of these appliances. I refer to the diminutive shaded pole motor that is most commonly used in this application. The 50/60 Hz line current which is drawn by these small motors range from between 0.25 A and 0.5 A. They are entirely undemanding in their control requirements merely requiring connection to the utility line in order to work. Their present switch connection is performed by the electro-mechanical timer/programmer which is fitted to the vast majority of these appliances.

The introduction of the ubiquitous microprocessor as the program controller will require that these small motors be switched by some interface device. An electro-mechanical relay would demand far too much drive from the output connection of the micro-controller itself. Opto-coupled solid state relays as shown in the circuit of Figure 9.9 can be driven directly by the microprocessor itself. The entry of this circuit is justified on the grounds that the power element in the solid-state relay which is depicted in the schematic uses unconventional MOS thyristor technology which is fabricated by at least two vendors.

10
Automotive Electronics

The electronic design engineer who has to design products for the automotive industry is faced with the task of having to take into account several conditions which are in some ways unique to this arm of the electronics industry. These may be briefly defined as:

(1) Ambient environment with respect to temperature, dust and humidity.
(2) Mechanical environment with respect to such mechanical considerations as shock and vibration.
(3) Transient excursions of the supply voltage.
(4) Legal considerations. Maximum allowable voltage drop for switches.
(5) Cost considerations.

When all of the ramifications of the five special conditions outlined above are taken into consideration, the task of the designer would initially appear to be untenable at best and impossible at worst. Any electronic circuitry would require components which could be considered as being better than MIL quality parts. Power switches should ideally have zero on resistance, be able to block exceedingly high voltages and cost almost nothing to purchase. My remark about parts having to conform with virtual MIL STDs in no exaggeration.

The environment within the engine compartment is as hostile as anything which could be considered for military and aerospace applications. Conditions can become so bad that non-hermetic relays must be protected within the passenger compartment if they are to operate with any reliability whatsoever. The ignition transistor, in particular, highlights almost all of the difficulties which must be faced.

The type of switch which is best suited for automotive applications will without exception be found to be a MOSFET for all circuits where the nominal battery voltage is either 12 or 24 V. The fundamental reason for this assertion is that 60 V BV_{DSS} parts are eminently suitable for the various functions. At this working voltage the MOSFET utilises silicon most efficiently.

In the case of the 12 V battery function which is the present-day level for cars, with the exception of certain circuit functions, the MOSFET must be able to withstand sustained high-energy voltage transients of approximately 48 V

(This is the latest internationally agreed level for the so-called load dump requirement.) In addition to the load dump specification, the semiconductor switch must be capable of withstanding a continuously applied voltage of 24 V. (This condition is termed as either the jump start or dock-side start condition and arises out of the possibility of the engine having to be started without a battery being permanently connected.)

In addition to the high-energy transient of 48 V and the sustained application of 24 V the semiconductor switch must also be able to withstand low-energy transients of 200 V or possibly as high as 300 V in some cases. This transient is sometimes referred to as the *ignition spike*. The voltage levels and the energy levels should be of little concern to designers of car electronics since the avalanche capability of most of today's MOSFETs is perfectly adequate for their self-protection.

If all of the foregoing voltage requirements were not enough to cause concern then the requirement for the switch to withstand a reverse battery connection with a simultaneous shorted load must also be taken into account. A brief outline of the requirement is given below.

Consider the consequences of a mechanic inadvertently connecting the battery incorrectly in a car which had sustained major front-end collision damage resulting in lamps having their housings severely damaged and with the holders being effective short-circuits. The resultant short-circuit current could constitute a fire hazard. The validity of this requirement may be questioned as being no worse than if the battery had been correctly connected. The supposition in this instance is that the quasi-intelligence surrounding the semiconductor switch would have over-current protection and would therefore not constitute a fire hazard. The same switch's internal circuitry would not be able to cope with the reversed polarity of the current.

The particular areas to which power electronics may be applied within the automotive industry are extremely diverse and may be classified, for the purposes of simplicity, into two broad categories as follows:

(1) Chassis electronics.
(2) Body electronics.

It is now pertinent to expand these two categories into their various constituent sub-categories.

10.1 CHASSIS ELECTRONICS

The broad category covers virtually all aspects related to the chassis and running gear of the car and covers diverse areas which may be sub-divided into lesser categories such as the complete power train (engine and transmission), brakes, steering and suspension. These lesser categories may now be studied in some detail.

Engine

Electronics will play an ever-important role in engine control because of the rising awareness of the car (and for that matter the bus and the truck) being one of the major sources of pollution. The use of electronics can reduce pollution significantly. Its main use for the engine will be in the inclusion within engine management.

The engine management system or Engine Management Unit (EMU) in most modern cars covers the two areas relating to fuel control and ignition.

Under the further sub-category of fuel control two further sub-categories are most prominent. These are the actual pressure feed (pump) of the fuel from the gasoline tank to the fuel inlet controller.

Fuel pump

The requirement for electronic control of the fuel pump or fuel feed arises out of the necessity to conserve fuel, at all times where possible.

With cars increasingly being fitted with fuel injection it becomes relatively simple to incorporate some form of fuel cut-off during engine braking or when the engine is operating on the over-run. The most easily achieved technique would be to use a solenoid valve which could be operated by an electro-mechanical relay. It is significantly more advantageous to control the velocity of the fuel itself (speed control of the pump) to allow for variables in the environment and to allow for rapid control during emergencies. These variables can be simplified to include fuel level in the tank, severe braking to avoid a collision, etc. It is accepted that the fuel injection system could cope with the foregoing variables, but rapid control of the fuel pump itself has the advantage of reducing the risk of fire during a collision. Fuel cut-off, if required, is extremely simple to incorporate.

This is easily achieved in principle in the circuit of Figure 10.1. MOSFET Q1 is driven from a fuel velocity controller which need be nothing more exotic than a simple pulse-width modulator with feed-back from a simple pressure sensor. The original reference to the PWM can easily be set by the ubiquitous micro processor of the EMU. Secondary switch Q2, a small BJT in this instance, removes gate drive to Q1 via D1. The collector of Q2 could be taken to turn on a brake switch P-channel MOSFET if rapid braking of the fuel pump is required. The P-channel MOSFET could in turn control a solenoid valve if required.

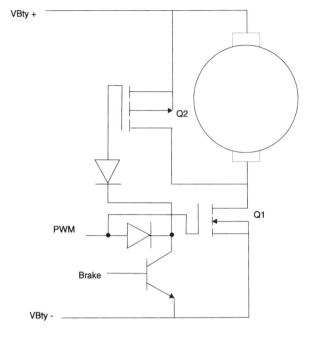

Figure 10.1 Fuel pump speed control

Fuel injection

Fuel injection is applicable to both gasoline and diesel powered cars and in the case of the former the application may be further sub-divided into *throttle body* (also referred to by some automotive personnel as electronic carburettor) and *manifold injection*.

The manifold form of injection can in turn be broken down into single-point and multi-point (simultaneous or sequential). If the preceding points were not enough to introduce confusion the reader should also bear in mind that there are variances in the concept as to how best to control the injection of fuel.

The schematics of Figure 10.2 illustrate just two approaches as to how best to control the injection function. In Figure 10.2(a) the injector solenoid is opened for however long the micro-processor may deem necessary and the solenoid is then allowed to close. The solenoid current in Figure 10.2(a) varies from zero to a value determined by the ratio of the battery voltage and the solenoid resistance. In the schematic of Figure 10.2(b) current is rippled just below the solenoid open level by switch-mode control of the gate of S1 during its closed state. When the solenoid is required to be open S2 is turned on by switch-mode control (and S1 off) for the period demanded by the micro-controller and the solenoid current is rippled just above the solenoid open level. The claimed advantage of Figure 10.2(b) is faster and more accurate control of the injection

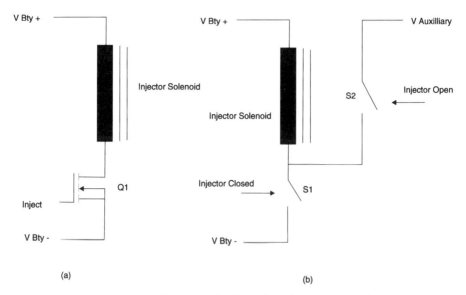

Figure 10.2 Conventional and switch-mode injector control

of fuel leading to the development of so-called lean-burn engines, which will in turn lead, it is claimed, to improvements in both fuel consumption and exhaust pollution. Accurate control of injection must lead to improvements in fuel consumption. However, the case for reductions in the level of pollutants is debatable. A more detailed schematic of Figure 10.2(b) is given in Figure 10.3. Operation of the circuit is as follows: initially the inject command is at logic 0 which sets the gate of the P-channel MGT Q1 high (therefore Q1 is off) and S1 (which may be a changeover analog switch) connects I_{ref1} to the reference input of the PWM IC.

The PWM switches Q2 to keep the current through L1 rippling below the value where the injector solenoid is only just closed. Solenoid current flows from the battery positive terminal through D1 and commutates from Q2 (when off) through D2. When the inject command goes to a logic state of 1 the gate of Q1 is taken low by the inverter stage and the inverter stage also sets the reference input to the PWM to I_{ref2} via S1. The upper side of the injector solenoid is connected to Supply 2 which is set to be greater than the battery positive potential. The current through L1 ramps up rapidly to the injector open value and the PWM ripples the current through L1 at this new open value until the inject command once more returns to logic 0. It is this current mode of control which gives the injection system its fast response. Although the circuit shown is for one solenoid it can readily be adapted to drive a multiple number of injector solenoids. A current mirror MOSFET is used in the position of Q2 in order to simplify the measurement of current flowing through the solenoid.

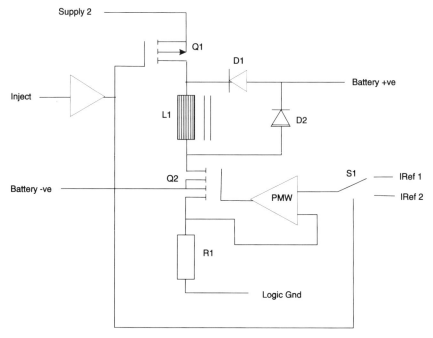

Figure 10.3 Simplified circuit of the precision fuel injection system

Electronic throttle control

It is not only the flow of fuel which can be controlled electronically. The rate of flow of air into the engine, whether by natural aspiration or compressor, super-charger or turbo-charger, must also be controlled. This is usually controlled by a restrictor, the butterfly. The concept of drive by wire becomes evermore a reality. The position of the butterfly is easily controlled by a stepper motor.

The power of the motor, in the car, is such that control can be performed by several of the proprietary power ICs on the market. The bus and truck function will require larger controllers.

The power controller for a two-phase stepper motor which could be used to control the position of the throttle butterfly in a bus or truck, where the peak current in any phase winding could be as high as 10 A, is depicted in Figure 10.4.

The four MOSFETs could be discrete parts or in turn could be four junctions which are mounted on a common insulator substrate. The maximum cold $R_{ds[on]}$ of each MOSFET should be no greater than 0.1 Ω. The overall cost of the circuit of Figure 10.4 including logic functions should be competitive with the power ICs which are available for lower power applications.

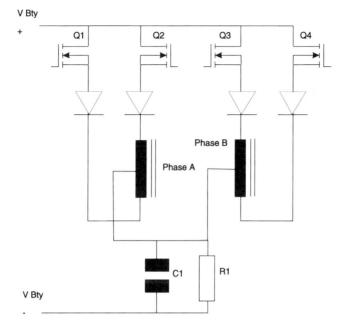

Figure 10.4 Two-phase stepper motor driver

Ignition

Since the ignition of diesel engines is mainly carried out by heating the air in the cylinder by compression, actual control and generation of ignition pulses will not be considered. In the case of multi-fuel engines ignition control is determined by the requirements of gasoline and kerosene and not by the diesel fuel constituent. The diesel engine which is predominantly used in cars is of the indirect variety. With this type of engine the fuel is injected into an auxiliary chamber in the cylinder which is usually pre-heated (possibly by Glo plugs) prior to starting. Electronic control of the Glo plug has not been available until recently. Control has been achieved by utilising electro-mechanical switches or relays. Reliability considerations have caused some manufacturers to investigate the use of solid-state switches. One type of automotive solid-state switch will be considered later in this chapter.

Gasoline fuelled engines require spark-plugs and high voltage pulse generators to fire the spark-plugs. The initial method used for the generation of sparks was the original Kettering coil which was used with the Kettering distributor. This system used a coil to store energy in the magnetic core of the coil by the build-up of current in the primary. At the time that the spark was required a contact breaker, actuated by a cam in the distributor, would open the primary circuit of the coil. The collapse of current in the primary and resultant collapse of the magnetic field would create a high voltage in the secondary of the coil. There were several disadvantages to this system of ignition spark generation. The major reason for its limited survival time was that electronic generators proved not to be really superior.

These limitations and disadvantages have been listed below and are as follows:

(1) Imprecise control of the timing and duration of the spark.
(2) Poor reliability due to wear of he actuator heel of the moving contact of the contact breaker. This wear also affected (1) above. Reliability and performance were also affected by arcing of the contact surfaces themselves.

The advent of the thyristor and the high voltage BJT made electronic control of spark generation possible. The initial technique was to charge a capacitor to between 300 and 400 V and to discharge the capacitor through a SCR into the primary of the standard coil. This technique, known as capacitor discharge ignition or CDI, gave certain benefits but also introduced limitations of its own and is now mainly confined to certain high-performance engines and since it advocates the use of thyristors will not be discussed further. The high voltage BJT and especially the Darlington derivative enabled the contact breaker of the original Kettering system to be replaced by a solid-state switch with consequent improvements to both reliability and performance.

The major improvements provided by the replacement of the contact breaker of Figure 10.5(a) with the BJT of Figure 10.5(b) was towards overall reliability. This was the result of allowing the original contact breaker to control a low current of some mili-ampères as against the several amperes controlled by the original contact breaker circuit. Wear on the actuator heel still proved to be the one limitation which required frequent adjustment, if overall performance was not to degrade at all.

The ultimate replacement of the contact breaker by Hall effect sensors or opto-couplers enabled performance to be kept at an optimum level without the need for continuous adjustment. The Darlington was not without limitations of its own. The reader's attention should be drawn to the presence of the Zener diode connected between collector and base. The Zener performs three functions:

(1) The switching locus of the BJT is kept inside the SOA curve of the device for the safety and reliability of the BJT.
(2) The $V_{CEO(SUS)}$ rating of the BJT must of necessity be greater than the Zener voltage.
(3) Automatic clamping of the collector voltage excursion prevented the coil from being over-volted in the unlikely event of accidental removal of the King lead from the distributor.

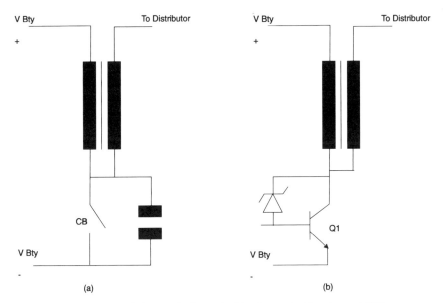

Figure 10.5 Inductive discharge ignition by contact breaker and BJT

The introduction of the MOSFET with its associated ruggedness enables the automotive designer to consider an alternative switch which should further contribute towards reliability. The fact that the high-power MOSFET has an avalanche capability means that the Zener diode (as used by the BJT) is not required. The reduction in component count is the first factor to contribute towards the improvement in reliability. The MOSFETs greater overall ruggedness should be the second factor. Removal of the Zener also means that the MOSFET solution should provide the additional benefit of reduced overall system cost if the Zener diode has not been monolithically integrated into the BJT structure.

Unfortunately the selection criteria of the MOSFET is not as simple as it would initially appear to be. The BV_{DSS} rating should be high enough to ensure the firing of fouled spark-plugs. The price that would normally be paid for this voltage rating is increased $R_{ds(on)}$, with a consequential increase in losses. It is, therefore, not surprising to find 400 V MOSFETs being considered in an application where 250 V parts would be perfectly adequate. This misunderstanding arises out of a misunderstanding related to the turns ratio (high voltage secondary to primary) of the ignition coil. The definition of the BV_{DSS} is given as:

$$BV_{DSS} = V_{EHT(O/C)}/N_{EFFECTIVE} \qquad (10.1)$$

where $V_{EHT(O/C)}$ is the desired open-circuit secondary voltage and $N_{EFFECTIVE}$ is the effective turns ratio of the coil.

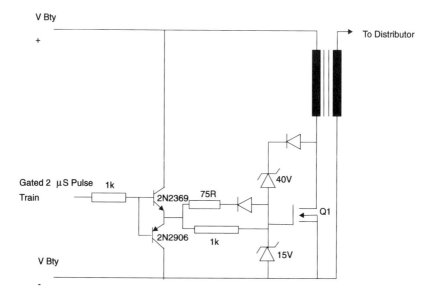

Figure 10.6 Multiple spark high frequency ignition system

There is a discrepancy between the effective turns ratio and the measured (or actual) turns ratio. The former is always higher than the latter and is affected by the Q of the coil secondary and might well be up to 50% higher. $V_{EHT(O/C)}$ should never be more than 35 kV since fatal break-down of the coil becomes possible and a figure of 30 kV is usually found to be perfectly adequate. If the equation for BV_{DSS} is used with an effective turns ratio of 105:1, not 100:1 as measured, then the MOSFET requirement is for 30 kV in a device with an $BV_{DS(AVALANCHE)}$ of 285 V. This avalanche voltage will be met with a device with a BV_{DSS} of 250.

The greatest potential improvement should be realised by changing the ignition generator to a high-frequency unit. Such a generator is shown in Figure 10.6.

Initially it would appear that there is only a small change to the circuit of the original Kettering type of spark generator. This is fundamentally true, with the exception that the high voltage coil has been changed to being a wide-band type which has been wound on a ferrite core and the MGT now requires a BV_{DSS} of only 60 V. The high voltage pulse train should be gated to provide a burst of pulses that last for no more than 1 ms.

The theoretical advantages of such an ignition generator are numerous and the major ones are as follows:

(1) Improved fuel consumption combined with significant improvements in performance. This arises out of the fact that more of the fuel within the

combustion chamber is burnt because of improvements made to the burning of the mixture. In an engine with a conventional generator, if the flame-front should extinguish for whatever reason, combustion will be incomplete. With the HF generator an ignition pulse should be available to re-ignite the unburnt mixture.
(2) An alternative to (1) would be to increase the ratio of air to fuel of the mixture, thereby making the engine one of the lean-burn variety. Optimisation of the mixture could conceivably lead to reductions in pollution.
(3) Engines which tend to be smoother in operation, namely. This has been demonstrated to be real.
(4) New type of coil increases the scope of utilising one coil to fire more than one cylinder which in turn ensures the elimination of the distributor.

Verification of the advantages highlighted above are beyond the scope of this book, but the claims have been included to demonstrate the potential of electronics to improve the performance of internal combustion engines.

Transmission

The requirements under this sub-heading tend to be more generalised into being mainly a requirement for solid-state switches for solenoid control, especially for the control of epicyclic gear trains as used in conventional automatic transmissions. This requirement is also true for the automatic electronic control of manual transmissions which are currently being developed.

Another recent development which also utilises solid-state switches for control of solenoids is *traction control.*

One of the methods which is used for this application is the sensing of wheel-spin in one or all of the driven wheel(s) and a cut-out to the ignition system or fuel injection system to reduce the power of the engine.

An alternative technique is to use the braking system of the vehicle in a manner where the brakes (to the driven wheel(s)) are applied in the event of the on-set of wheel-spin. Since this feature can be easily incorporated into the car's braking system as an additional feature the salient points will be covered under that sub-heading.

Brakes

Conventional braking systems in cars have been honed to a fine art with the use of hydraulic actuation of calipers, for disc brakes, and/or shoes in the vast majority of vehicles.

Anti-skid braking systems (abbreviated to ABS) utilise wheel lock electronic sensors which activate the closure of hydraulic solenoid valves to remove the application of pressure in the hydraulic circuit in the event of wheel lock-up being imminent. The solenoid valves are deactivated when wheel lock-up is not likely (or after the passage of a predetermined time interval). The power electronics requirement for ABS reduces to solid-state switches for activation of the solenoid valves.

An extension of ABS is traction control. In this instance it would be a requirement for the brakes to be applied through a secondary pressure-fed hydraulic circuit with actuation of the brakes being solenoid controlled to reduce the tractive effort being applied through the driven wheel(s).

Suspension

Modern cars maybe equipped with one of a variety of suspension types. These may be of the self-levelling type which in turn may be either active or of the passive hydro-pneumatic type. Alternatively the type of suspension could well be totally passive while a third variety belongs in the classification of being active whilst embracing more than mere self-levelling within its functional domain. The activities which may fall within the fully active domain could cover the following two suggestions:

(a) Ride height.
(b) Ride compliance (normal or sporting are examples).

Ride compliance could in turn cover many variances such as soft, medium or firm within the normal context and the sporting driver could well be given the option of controlling the response time which could in turn affect the overall handling and road-holding.

In theory the active suspension medium could be one that is neither hydraulic or pneumatic nor a combination of the first two. It is feasible that an exotic medium such as magnetic levitation could be countenanced although this is unlikely in the foreseeable future on the grounds of cost. The almost universal spring medium will therefore be either hydraulic or pneumatic or a combination of the two. The power electronic functions will therefore inevitably reduce to being solenoid controllers.

The final applications area for power electronics inside the domain of chassis electronics is that of steering. Almost all vehicles which offer the option of power assistance to the steering function make use of hydraulic pressure derived from an engine-driven pump. It is to the credit of the mechanical engineers that the system works as well as it does and with the reliability which has been achieved lately. It is inevitable that this function will ultimately fall inside the scope of power electronics. The advantages of an electronic power steering system over a hydraulic may be quantified as follows:

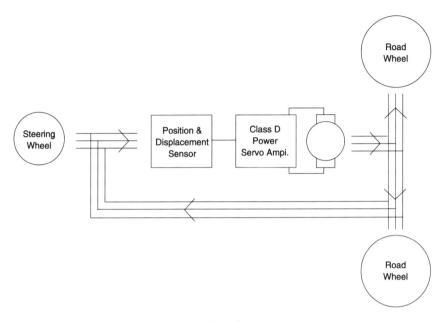

Figure 10.7 Block schematic of an electronic power steering system

(1) No performance sapping hydraulic pump to be driven by the engine. Power will instead be drawn from the battery as and when required.
(2) Power steering systems now easily become a retrofit option which could easily be carried out by the owner.
(3) Inventory held by the manufacturer would reduce to a single part per model instead of the present two items.

The three advantages above are only some of the benefits which would be arrived at by the adoption of an electronic power steering system.

The circuit of Figure 10.7 uses a full bridge since four-quadrant control of the motor is necessary.

10.2 BODY ELECTRONICS

The power electronics designer should pause before considering all of the functions which may be included under this heading and consider the nett effect upon other aspects of car electronics.

Power for all functions which are powered from the electrical source of the car requires an increasing power capability of the alternator and of the size and complexity of the *wiring harness* within the car. The dramatic increase in electrical functions has ensured that the mass and bulk of the wiring harness has become critical in certain classes of luxury cars. The ultimate solution to this problem would be the installation of a multiplexed (MUX) electrical system within these cars. Such a system would require that all solid-state power switches become micro-processor controllable with built-in diagnostics within the switch to indicate problems which may arise owing to failure of the load.

These SMART switches simplify the task of the engineer into becoming more of a systems-oriented person. It therefore implies that for the purpose of this book the MUX concept and SMART switches will be neglected. Instead we will turn our attention to the enormous diversity of power electronic functions which are to be found under the heading of body electronics. A large number of the functions fall under the category of motor control while the majority of the remainder may be classified as general load switching.

Motor control

Four-quadrant controllers of brushed d.c. motors may be contained within the following functions: (1) memory seat modules, (2) door mirror positioners, and (3) power window lifters. The power circuit would follow the principle as contained in the schematic for the power steering function. It would be only the logic and small-signal function of the servo loop which could differ. In certain circumstances these motors and their associated controllers could conceivably be replaced by permanent magnet motors.

Detailed descriptions of different types of motor controllers are to be found in the preceding chapter.

Simple motor controllers, using a single switch, will be found to perform many functions within the car. The applications which are most common may be listed as follows: windshield washer, pop-up headlamps, retractable radio antenna, radiator cooling fan and interior fan. The list given here is representative of the number of simple motor controllers which may be found in cars.

As stated in the previous paragraph the type of circuit which controls these motors would involve the use of a single solid-state switch which could be either *low side* or *high side* configured. The terminology of these switches can be found in the schematic of Figure 10.8. It is worth noting that it is of little importance whether the switches are pulse-width modulated or not. It is the application and not the configuration of the switch which will determine the use of PWM in the circuit.

Factions within the automotive industry will advocate one or other of the configurations as being superior. Reasons for their choice are frequently cited as reduction in both corrosion and leakage current (a potential drain on the battery of a car which has been left unattended, e.g. in an airport car park, for some days. There is little or no evidence to support these hypotheses. Another reason which is frequently given as favouring the high-side switch is fail-safe protection in the event of the non-battery pole of the switch being shorted to

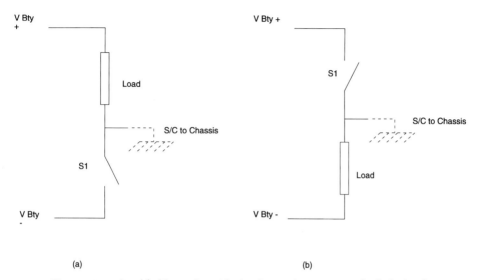

Figure 10.8 Simplified low-side and high-side switches. (a) Low-side, (b) high-side

the chassis. This point is perfectly valid for certain high-current applications which could endanger life, e.g. starter solenoid or radiator fan. The point is not valid in the case of lamps or antenna motors.

The configuration of the high-side switch could easily be accomplished with a P-channel MOSFET. Unfortunately the area of silicon required to make a P-channel device is usually double that of the area of a N-channel part with the same efficiency. Economy therefore dictates the use of N-channel MOSFETs. Unfortunately level shifting and an auxiliary supply 5 V higher than the battery voltage would be required to ensure full enhancement of the switch. (The 5 V requirement is for logic level MOSFETs, standard devices would require an auxiliary supply of 10 V.) If overload protection of the switch is required then the difficulties are magnified and could possibly make the overall cost unacceptable, especially for high-current applications which require low ohmic drops. The control switch for a car's headlamp is an excellent example. The cold resistance of the lamp's filament usually implies a current surge equal to five or six times the normal running current. Frequently legal requirements related to a particular nationality may require that no more than 0.25 V be dropped across the switch. The current through a pair of 60 W lamps would equate to 10 A which would require a MOSFET with a heated $R_{ds[on]}$ of 25 mΩ. The cold $R_{ds[on]}$ for the MOSFET in this instance could well be as low as 10–12 mΩ. Monolithic protected switches of this calibre may well be difficult to obtain and the unit cost of such a part could prove to be prohibitive.

The reader could be forgiven for coming to the wrong conclusion that this particular switch is best fulfilled by electro-mechanical means.

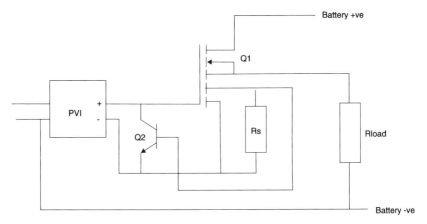

Figure 10.9 Simple over-load protected high-side switch

It is fortunate that such a high-side switch can easily be configured from discrete components which are freely available. The principle has already been given earlier in the book but its inclusion here will not be taken amiss.

The circuit of Figure 10.9 shows an overload protected high-side switch. It should be noted that in this instance it is imperative that a current mirror MOSFET be used, since a series sense resistor would make the total resistance of switch untenable. The secret of this switch is a unique component known as *photo-voltaic isolator* (PVI). This IC combines the virtue of being an opto-coupler and a floating supply. It is important for the reader to be aware that high-frequency switching is not one of the properties of this switch. This is the limitation imposed by the output current capability of the PVI. The current limit BJT needs to be an ultra-low leakage device which does not latch-up with the extremely small collector current which would flow through the part during current limit operation of the main power MOSFET. This considerably simplifies the definition of the sense resistor R_S.

$$I_{D(Q1)} = V_{BTY}/(R_{ds[on]Q1} + R_{LOAD}) \qquad (10.2)$$

Because $I_{be(Q2)}$ is so small since $I_{C(Q2)}$ would in turn be small and the gain of Q2 can be ignored, it is relevant to neglect any part played by the base current of Q2 in determining the value of R_S.

$$R_S = V_{BE(Q2)}r/[I_{D(Q1)}] \qquad (10.3)$$

where r is the sense ratio of Q1.

Substituting for $I_{D(Q1)}$ gives:

$$R_S = V_{BE(Q2)}r(R_{ds[on]Q1} + R_{LOAD})/V_{BTY} \qquad (10.4)$$

The circuit of Figure 10.9 is admirably suited to switching lamp loads, where the current limit function is now gainfully used to limit the *in-rush current* of the cold lamp which is frequently a cause of failure in filament lamps. The actual failure mechanism caused by in-rush current is not only of the single variety. The mechanisms range from magnetostriction to shock and thermal fatigue and may possibly also encompass a failure in the halogen cycle, if the lamp is of the quartz halogen type.

11
Electronic Ballasts

Electronic ballast deserves special consideration since it is usually employed for the driving of a completely unconventional load. The complex nature of the load is frequently alien to the engineer who has to develop a ballast circuit for the first time.

Imagine the consternation of the engineer when initial testing of the prototype ballast, with an uncommonly familiar fluorescent lamp as the load, results in an unexpected display of pyrotechnics on the part of the ballast and disappointment on the part of the individual as a result of failure — not usually appreciated by the neighbouring individuals within the laboratory or similar establishment. If the perpetrator of the said pyrotechnics was questioned as to their knowledge of Townsend discharge, abnormal glow and acoustic arc resonance the chances would be that these titles would mean little to the unfortunate individual. This lack of knowledge, or understanding, coupled with the earlier failure of the prototype ballast is certain to raise little or no enthusiasm for further experimentation.

A full and meaningful account of the three titles and others related to discharge lamps is not possible within this book. They are merely presented to illustrate the specialised nature of the load presented by the discharge lamp to the electronic ballast.

Many of the lamp characteristics should be taken as read with the full understanding that there is a negative impedance characteristic of the lamp during operation in the arc region (the desired region of operation.) The reader should also be aware of the possible need for differing wave-shapes which may be used to drive the tube.

The negative impedance characteristic is the reason for the existence of the *magnetic ballast* which has been used so successfully and for so long. The ballast's sole *raison d'être* is to behave as a current generator for the tube after it has struck. The need to replace the magnetic ballast arises from the quest for higher efficiency, longer ballast life and the elimination of tiresome 100/120 Hz flicker in the case of the low pressure fluorescent lamp, with the added benefit that domestic fluorescent lamps and those used in commercial office establishments can be relatively easily dimmed with the use of electronic ballasts. This last feature is extremely difficult to achieve with conventional magnetic ballast,

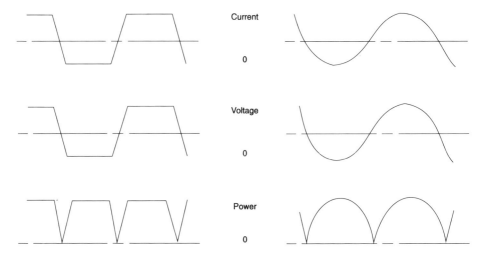

Figure 11.1 Idealised current, voltage and power wave-forms. (A) square, (B) sine

especially when they are used with the low pressure fluorescent lamp and with most of the high pressure discharge lamps. This last statement is found to also apply to certain metal halide lamps.

Having covered the introductory preamble related to the electronic ballast, it is now necessary to define some of the prerequisites for the electronic ballast. These are usually categorised as:

(1) Is the current wave shape as delivered to the lamp sinusoidal or square?
(2) Frequency of the current wave shape.
(3) Supply constraints of the ballast.

The negative impedance characteristic of the lamp has determined that it should ideally be driven by a current source and the relative merits of the two most frequently used wave shapes need to be considered.

From the waveforms in Figure 11.1 it should be apparent that the square wave of current provides the optimum drive to the gas discharge lamp. A simple test with two differing ballasts (one sine wave and the other square wave) will further demonstrate that the lamp efficiency is marginally improved with a square wave drive. The equipment required for the test is confined to (a) an electrical dynamometer and (b) a photometer.

The two measurements which should be the result of the tests are input power for the two lamps and the respective light output for the measured input power. The overall efficiency is the ratio of light output and input power and is expressed thus:

$$\text{Lamp efficiency} = (100 \times \text{light output}/\text{input power}) \qquad (11.1)$$

11.1 EFFECTS OF HARMONIC CONTENT OF DRIVE CURRENT WAVEFORM

where lamp efficiency should be truly denoted by a figure of merit and light output is measured as lumens/watt.

It is now necessary to define the optimum frequency at which the ballast should operate.

Once again it can be demonstrated that for increasing frequency up to approximately 75–80 kHz there is a correspondingly modest increase in lamp emission. It is pertinent to point out that the difference in lamp emission from 50 Hz to 80 kHz will not be greater than 2–5%. It is debatable if this information is necessary. It is given in case the occasion arises where maximum efficiency is the only factor to be considered.

There are perfectly valid reasons why both the current wave shape and the frequency are of such significance. First, it should be remembered that the harmonic content of a square wave is considerably higher than the harmonic content for a sine wave of a similar frequency which corresponds to the fundamental of the square wave. The effects of the harmonic content are profound to say the least.

11.1 EFFECTS OF HARMONIC CONTENT OF DRIVE CURRENT WAVEFORM

The effects are basically twofold and are confined to conducted and radiated interference.

The effect on conducted interference can be easily rectified. A simple line filter should be adequate to remove almost all pollutants which may be fed back to the line. The line filter itself may prove to be unnecessary if power factor correction is employed in maintaining the input line current as close to a sine wave as may be required.

Radiated interference and its consequences are much more insidious. There are two sources and they should be treated separately. First, there are the leads which connect the ballast to the lamp. These leads could easily prove to be antennae for the harmonic content of the output waveform. Their influence can be reduced with the careful use of screened leads or field cancellation (the use of a twisted pair of conductors). The disadvantage of the use of screened leads, which will provide the most attenuation, is an escalation in the cost of the complete system. The use of a twisted pair of leads therefore becomes mandatory. Of greater significance will be the lamp itself. It too will function quite efficiently as an antenna and reducing these emissions will prove to be nearly impossible.

The conclusion to the choice of wave shape therefore is that, although square wave current drive is preferable to a sine wave of current, its implementation will be limited to those applications where interference has a lower priority than overall system efficiency.

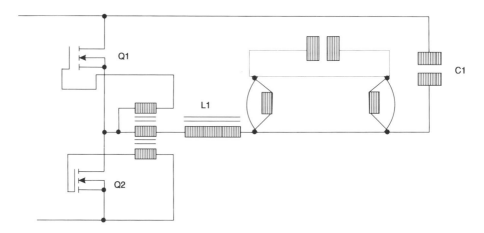

Figure 11.2 Schematic of usual sine wave electronic ballast

The circuits which may be used to generate the two current waveforms are themselves shrouded by misunderstanding and fantasy.

Examination of the circuit which has been most widely adopted for generation of sine wave currents (as depicted in the schematic in Figure 11.2) soon establishes it as belonging to the family of series resonant topologies.

The use of the half-bridge topology is of itself quite evolutionary, since this circuit can, relatively easily be made to be self-oscillating. The circuit itself has the additional distinction of operating at a level of efficiency which can be regarded as being admirable. The use of the current transformer to achieve the regenerative gate drive signals means that the circuit can be easily adapted to a dimmable application. The most widely used technique for dimming is to utilise the band-pass filter characteristic of the series resonant converter and to increase the frequency of operation above resonance.

There is a very good reason for increasing the frequency above resonance. At resonance the impedance presented to the totem-pole driver is resistive in nature. This is, however, no longer the case for frequencies above and below resonance. Above resonance the impedance presented to the totem-pole is predominantly inductive and poses no problem to the switches since the current to the load lags the applied voltage. At frequencies below resonance the impedance to the totem-pole unfortunately is predominantly capacitive and the current leads the applied voltage. This situation now creates the problems related to t_{rr} of the parasitic diode in MOSFETs as discussed in Chapter 9.

The necessary alterations to the dimmable ballast are shown in Figure 11.3. It should be noted that the alterations are entirely confined to the drive transformer. The additional circuitry uses a 555 timer in the astable mode (as a clock generator) and the D type flip-flop is a dual 50% duty cycle generator

11.1 EFFECTS OF HARMONIC CONTENT OF DRIVE CURRENT WAVEFORM

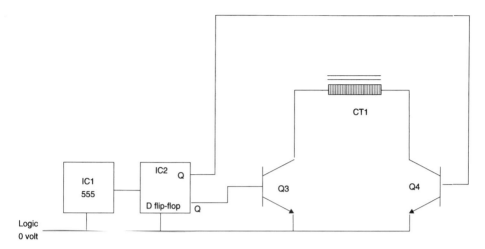

Figure 11.3 Frequency control of dimmable ballast

with a minute overlap period. The two Darlington transistors provide the short-circuit reset to the core of the drive transformer, which reverses the drive upon the removal of the short-circuit. Frequency modulation utilises the filter type of frequency response of the resonant circuit to achieve dimming. It is necessary to operate at, or slightly above, resonance for maximum light output. The frequency should be increased to achieve the desired level of dimming.

The initial design of the resonant circuit in Figure 11.2 is simple in the extreme. The major components are L1 and C1. These two elements comprise the resonant circuit itself. The purpose of C2 is to prevent the lamp attempting to behave as a rectifier. (The reader is advised that fluorescent lamps are perfectly capable of operating with a d.c. current. Careful examination would show that current controlled d.c./d.c. conversion would not be entirely cost-effective in this application.)

The component design should proceed as follows:

(1) Define V_{DD}. This d.c. value is nominally equal to $2^{0.5} V_{LINE(r.m.s.)}$.
(2) From lamp data obtain the operating current i_L for the lamp. It should be remembered that this is a r.m.s. value.
(3) From lamp data obtain the operating voltage v_L for the lamp. Again note that this is a r.m.s. value. If the operating voltage is not known it is reasonable to assume a value of 110 V.
(4) Use $L < V_{DD}/[2(2f_o i_L)]$ to determine the value of L1.
(5) Use $C < i_L/(V_{DD} f_o)$ to determine the value of C1.
(6) Finally, use $V_{DD}/i_L = [L1/(C1/C2)]^{0.5}$ to find the value of C2.
(7) The choice of core for CT1 should be based upon the volt-second saturation requirements of the core. The voltage will be the secondary voltage which

must be greater than the sum of Zener diode voltage which should not be greater than 10 V for reasons of reliability and the voltage across the resistor in series with the gate. The value of this resistor should be used for the purposes of fine tuning the frequency.

The value for v_L from (3) above would lead to the assumption that a fluorescent tube with an electronic ballast could be operated off a rectified 110 V line. This assumption is found to be perfectly valid.

The inexperienced user may query how starting voltages of 1.5–2 kV can be accommodated in a circuit where the maximum supply voltage may be 370 V (rectified high UK line). The user is reminded that the voltages appearing across L1 and C1 are in quadrature. L1 alone defines the current which is forced through the tube prior to the tube striking. The quadrature voltages ensure that there is always sufficient voltage to cause the tube to strike. This spike of voltage in no way endangers the MOSFETs themselves, while the high value of striking current is usually provided for by the MOSFET's overload capability.

Operation of two tubes from a single ballast is perfectly feasible in the ballast of Figure 11.2 provided that there are two separate resonant circuits, each with its own inductor and capacitor. This capability is illustrated in the schematic of Figure 11.4. The cross-connection of the two capacitors is a refinement which ensures that removal of one of the tubes automatically ensures that the remaining tube will fail to operate.

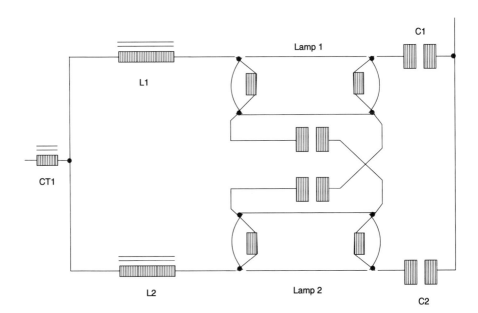

Figure 11.4 Dual lamp ballast

11.1 EFFECTS OF HARMONIC CONTENT OF DRIVE CURRENT WAVEFORM

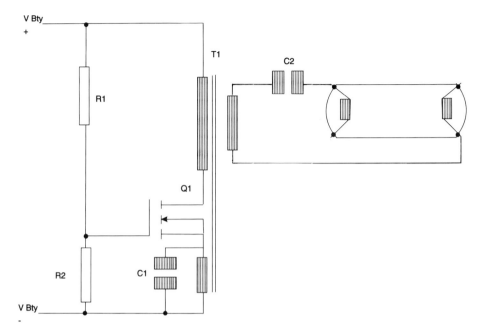

Figure 11.5 Schematic of simple electronic sine-wave ballast

A single switch variant of this ballast would be equally simple to design and fabricate. Its suitability for this application has not been very carefully explored. One of the objections which is frequently cited is that the single switch circuit requires the use of a transformer. This argument can be countered quite easily. The half-bridge of Figure 11.2 uses a magnetic component of similar volt-second saturation capability as the transformer itself. Another objection is that the circuit is difficult to be made to self-oscillate. This is again quite fallacious. A third objection, which is not proven, cites efficiency as being relatively poor. The only real argument against the adoption of this circuit is that it does not lend itself to easy adaptation for a dimmable application. The point which is overlooked is that simple emergency lighting types of units are not meant to be dimmable.

The fact that the circuit utilises only one switch should be the catalyst to promote further investigation as to its capabilities. The two beneficiaries for this reduction in component count are cost and reliability.

Reliability is a function of component count. Therefore a reduction in component count must make a circuit more reliable. Assuming, of course, that all other factors remained equal.

The proof, if such were required, would be an examination of portable fluorescent lamp sets for camping and other such activities. The ballast in this instance is nearly always a single switch sine-wave converter. A simplified schematic for such a ballast circuit is demonstrated in Figure 11.5. There is

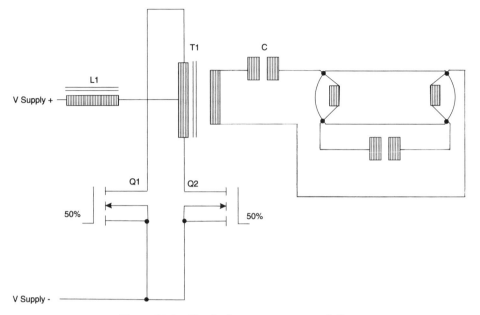

Figure 11.6 Circuit of square-wave current ballast

unfortunately one drawback to this simple circuit. The inherent dearth of tube heater current results in blackening of the tube wall near to the tube ends, somewhat reducing useful tube life. There are some who would contribute to the school of thought that the reduction in overall tube life is a relatively small penalty and since these lamps will be used relatively infrequently the meaning is that overall reliability does not manifest itself as being so severe a limitation as might have been first envisaged. It should be pointed out that the circuit of Figure 11.5 is better suited to using a BJT for the simple reason that this device lends itself to self-oscillatory operation using a blocking oscillator topology. It is considerably more difficult to use an MGT in this type of circuit and the additional cost of drive circuitry and the cost of providing power for the drive circuitry may prove to be unacceptable.

A circuit which may be adapted for use as a ballast for square-wave current generation is given in Figure 11.6. Experienced designers will recognise the converter circuit as being a current-fed symmetric push–pull.

There are those who will claim that this type of circuit cannot be realistically considered owing to the prohibitive cost and will again cite the extraneous components to support their complaints. Close scrutiny reveals that almost all of the criticism is unfounded.

It should be noted that the similar counter should be applied to the criticism concerning the use of the transformer and to the capacitor C which is connected in series with the tube and the transformer secondary. This capacitor, as stated

11.1 EFFECTS OF HARMONIC CONTENT OF DRIVE CURRENT WAVEFORM   177

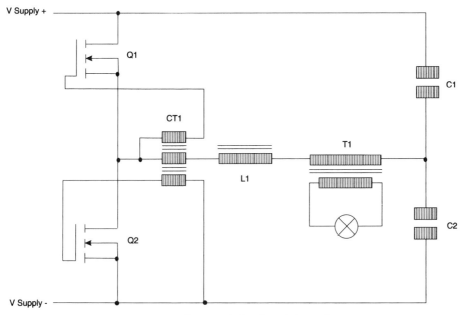

Figure 11.7   Electronic ballast for a halogen lamp

previously, prevents the tube from working as a rectifier and thereby causing the core of the transformer to saturate.

All of the ballasts which have been covered so far relate to gas discharge lamps which have a negative impedance characteristic over the region in which they are required to operate. More recently ballasts have been required to drive small quartz halogen lamps from utility lines. The application is unusual in that the equipment/appliance is a reading lamp which has an intense white light output of modest power. The lamp which fulfils all of the requisite demands is a low power halogen lamp. Unfortunately these lamps have voltage ratings of 12/24 V. Ballasting such a lamp purely with passive components is extremely difficult. The simple 50/60 Hz line transformer is ruled out owing to the overall mass.

Since this reading lamp does not normally have cost restraints placed upon it electronic ballasting has been adopted. Those who have witnessed the sale of these lamps will agree that they are priced for sale at the top end of the market.

Soft-start operation, a term relating to the gradual increase in load current after initial switch-on, can be easily incorporated to allow for the lamp's cold resistance. A simple Buck converter type of ballast, although being cost-effective, is ruled out on the grounds of safety — the simple Buck converter normally has no isolation transformer. Inadvertent lamp replacement with power applied exposes the person to line voltages and the hazard of electric shock.

178   ELECTRONIC BALLASTS

In the chapter related to motor controllers the question of compliance with International Safety Regulations was broached. It is for this very reason that the simple Buck converter battery charger cannot be used. One of the favoured circuits for this application is an adaptation of the ballast circuit of Figure 11.2. The adapted circuit is given in the schematic in Figure 11.7. It is apparent that no reservoir capacitor has been included since 100 Hz ripple has no effect upon the reader, since flicker in the light output has no adverse effect. The inductor L1 is required at switch-on to limit the in-rush current of the lamp. Although the load is fed with a triangular wave of current, radiation of the harmonic content does not pose too serious a problem. The size of the filament lamp is question makes it a relatively poor antenna regarding radiated EMI

12
Audio Amplification

One of the earliest areas of application for power transistors has been the audio power amplifier. This usage arose out of the realisation that the heat generated by the thermionic tube, even in non-power stages, was unacceptable and resulted in amplifiers of relatively modest output power levels. Once again it was the BJT which was first adopted for use in this field. This is logical since from a chronological standpoint the BJT arrived on the scene much earlier than the MOSFET.

In consideration of early power amplifier design and the subsequent use of BJTs it is to the eternal credit of the earliest transistorised power amplifiers (and more so to their designers) that the performance turned out to be as good as it was. The designers of these early power amplifiers made non-linear elements work in a linear mode to extremely good effect.

It is equally impressive that bipolar integrated circuits have been fabricated to operate as power audio amplifiers, of reasonable output power, within In Car Entertainment (ICE) applications, despite having to work with lateral PNP power transistors which are extremely slow, when comparison is made with the vertically fabricated NPN in the integrated circuit. The speed constraint arises out of the absolute necessity to fabricate the PNP as a lateral structure. This apparent asymmetry in the two types of output transistor in the IC has had no deleterious effect upon the performance of the integrated circuit power amplifier.

Designers of audio amplifiers have been attracted to the MOSFET and more recently to the possible inclusion of the IGT, although this may be regarded as a retrograde step, as a possible amplifier element, owing to the linear nature of its transfer characteristic. The performance of these MOSFET power amplifiers have been exceedingly well received.

Modern musical tastes and trends has resulted in an ever increasing demand for power audio amplifiers of ever greater output power capability. Amplifiers with output capabilities of 500 W per channel are relatively commonplace in the field of car booster amplifiers. Public address and rock concert amplifiers are not uncommon in requiring several kilowatts of output power. This quest for ever increasing output power levels has led to the investigation of parallel connecting ever greater numbers of MOSFETs, in order to achieve the necessary output power. Parallel connection of MOSFETs in linear applications pose no real problems for devices which have been specifically designed for the application. However, some care is advocated when general purpose vertical structure devices are considered.

Certain designers have started to investigate the use of switched-mode techniques within the audio amplifier as a means of reducing the huge quantities of heat which is generated within the amplifier. These switched-mode amplifiers on occasion are referred to as digital amplifiers, a term which is not strictly true since the loudspeaker still requires an analog signal of one sort or another in order to create the necessary acoustic pressure wave.

One of the unfortunate results of the investigation into switched-mode techniques, especially in the development of the car booster amplifier, has been the development of separate SMPS (switched-mode power supply) and pulse-width modulated output inverter. It will be demonstrated later in this chapter that the optimum degree of amplification results from a combination of SMPS and amplifier within a single circuit topology. I have used the term 'ampliverter' (an abbreviation arrived at from the front and rear portions of amplifying converter) for this particularly novel circuit idea.

The topics which will be covered within this chapter are:

(1) The parallel connection of MOSFET and any difficulties which may arise out of this type of MOSFET usage.
(2) PWM inverters and the body-drain diode.
(3) The ampliverter.

12.1 PARALLEL OPERATION OF MOSFET IN THE LINEAR REGION OF OPERATION

One of the selling features of the vast majority of MOSFETs has been the negative temperature coefficient of current positive temperature coefficient of resistance. This feature does not extend over the full operating range of drain currents and for all levels of enhancement. A glance at the relevant curves within manufacturers' data will demonstrate that below a certain value of drain current and for certain values of V_{GS} the opposite is true. In this region of operation the MOSFET has a negative temperature coefficient of resistance and may be regarded as having virtually all of the propensity of the BJT towards *thermal runaway*. I should point out that this is not completely true for all MOSFETs. There are devices in the market-place which have been specially fabricated for the audio amplifier market and these devices do not exhibit the feature I have described.

The last paragraph has been used to draw attention to the potential of the MOSFET amplifier proving to be unreliable when all of the earlier marketing propaganda may have led the unwary user into believing that the MOSFET had no apparent limitations.

12.1 PARALLEL OPERATION OF MOSFET

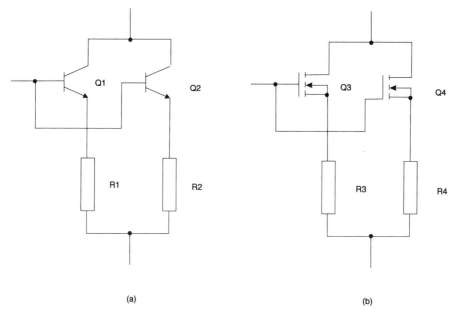

Figure 12.1 Schematic demonstrating emitter source ballsting

It is possible to force parallel-connected MOSFETs operating in their linear region to share current, just as it is possible to force the devices to share when they are parallelled in a switching application. The technique of forcing which is advised here for absolute reliability is to apply negative feed-back around the individual MOSFET as is done with the BJT.

There are mainly two techniques to be considered as means of applying negative feed-back. These may be stated as being:

(1) Source ballasting (the term is used to indicate the similarity to the use of emitter ballasting which is the normal method of ensuring BJTs shared current) when parallel connected.
(2) Gate control.

Source ballasting

There is nothing mystical about the term used here to describe the technique, being advocated, for the application of negative feed-back as a means of ensuring current sharing. The application of this particular approach is demonstrated in the schematic of Figure 12.1.

There are some who will state that the approach, illustrated here, is not the only one which is valid. It will be claimed that the technique of emitter/source ballasting can be demonstrated as being very inefficient and that greater efficiency can be obtained by the use of base ballasting. In reality, efficiency and all of its connotations is not in question at this point. The argument between base/gate or emitter/source ballasting concerns the safety of the power semiconductor and its ultimate long-term reliability. Since the reliability claims of base/gate ballasting can be shown at best to be questionable no further discussions related to its application will be entered into.

In the schematic of Figure 12.1 both versions: (a) for the BJT and (b) for the MOSFET, fulfil the same task for both types of devices. Those members of the power electronic fraternity who maintain that base ballasting can be demonstrated as being equally effective as emitter ballasting will find that their arguments are ill-founded.

In Figure 12.1(a) the values of R1 and R2 may be found by the judicious use of the following approach:

(1) Assume V_{be1} equates to V_{be2} which in turn may be approximated to 0.7 V in full data for the preferred transistor is not available.
(2) Using:

$$I_{C1} \times R1 \geqslant V_{be(max)} \leqslant I_{C2} \times R2 \qquad (12.1)$$

calculating the value of R1 and therefore R2 is now quite straightforward.

If unequal values for I_{C1} and I_{C2} are inserted into Figure 12.1 it will become apparent that the resulting inequality will cause the transistor with the greater value of collector current to become less forward-biased until a state of equilibrium is again in force. It should be noticed that the current gain of the two transistors is never considered. When base ballasting is attempted then both base current and h_{FE1} and h_{FE2} also enter into the equation when calculating the values of collector current for the two transistors. (The term base ballasting refers to the connection of resistors in series with the base of the BJT.)

It should be noted that when R1 and R2 are calculated using equation (12.1) the optimum amount of current sharing will be achieved. Unfortunately, this optimisation will be achieved at the expense of increased dissipation in R1 and R2. Perfectly acceptable results can be achieved if the value of V_{be} (inserted into the equation) was made to be 0.7–1.0 V as suggested.

The example of emitter ballasting has been presented in order to introduce the concept of source ballasting. The approach which is advocated as the means of calculating the values of R3 and R4 is as follows:

(1) Assume $V_{GS(th)1}$ equal to $V_{GS(th)2}$ and equal to 2 V if no suitable data for $V_{GS(th)min}$ is available.
(2) Using

$$I_{D3} \times R3 \geqslant V_{GS(th)min} \leqslant I_{D4} \times R4 \qquad (12.2)$$

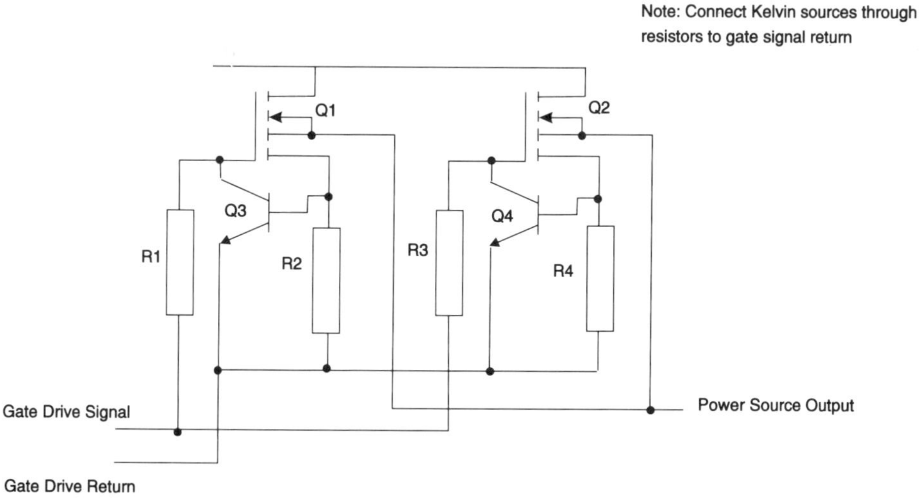

12.1 PARALLEL OPERATION OF MOSFET

Note: Connect Kelvin sources through resistors to gate signal return

Figure 12.2 Negative feed-back with current-sensing DMOSFETs

Just as it was demonstrated that the values of emitter ballast resistors calculated by the use of equation (12.1) result in the optimum degree of current sharing in the two BJTs, it can be demonstrated with equal validity that the degree of current sharing in the two MOSFETs will be optimised by the use of equation (12.2). It will be found that the penalty for optimum current sharing is considerably greater than optimised dissipation for the BJT. The unfortunate consequence of this increased tendency in dissipation is that designers of MOSFET amplifiers have tended to take risks with the level of source ballasting used.

Gate Control

The technique which will be demonstrated in this section will be seen as offering the greatest return in terms of efficiency, since no large values of resistance have been included in the power circuit of the source. All of the current measurement and control is performed at low power levels with the use of mirror or current-sensing MOSFETs. If used correctly it can be seen as providing the ultimate degree of current sharing, but at the unfortunate penalty of high unit cost.

The circuit of Figure 12.2 illustrates a simplified method of measuring the drain currents of the two MOSFETs and controlling their individual current-limit levels. If the current-limit levels are accurately defined to reflect dissipated power, then it will be seen as providing perfect safety but at the expense of poor sharing. The acceptance of the complete lack of 'lossy' components may persuade the adoption of such a circuit as being the best compromise involving safety without excessive heat-loss and at a price which may be regarded as being perfectly affordable.

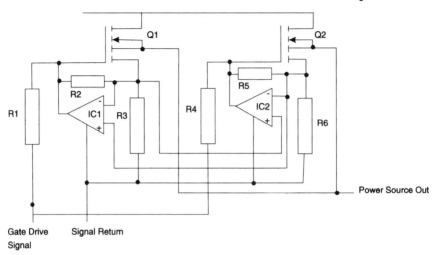

Figure 12.3 Total control of current sharing

The more ideal solution for this type of circuit concept is illustrated in the schematic of Figure 12.3.

The two-sense MOSFETs, in the circuit at left, have their two-sense outputs cross-connected to voltage comparators. Each comparator output is in turn connected to its corresponding gate circuit to ensure that a condition of equilibrium is maintained at all times within the drain currents of the two MOSFETs.

It should be appreciated that the circuit of Figure 12.3 does not constitute the most cost-effective solution. It merely demonstrates that it is almost possible to achieve perfect current sharing within MOSFETs which are operated in their linear regions.

In most linear power amplifiers a cost-sensitive compromise is arrived at where the control of current-sharing does not render a totally indestructible output stage. The power stage is designed to be virtually fault-free for most conditions which could arise in use and not for all possibilities.

Several other problems are associated with audio power amplifiers which are not all related to the power semiconductors. These, such as audio rectification, motor-boating, etc., are mentioned out of interest, but will not be covered.

Motor-boating results out of poor or inadequate decoupling and/or filtering of the power supply of the amplifier. The increasing demands of power output places greater emphasis upon the power supply, especially in the case of the car booster.

The primary power source of the car booster is the battery. The voltage of the battery is totally insufficient for the power levels of these amplifiers. It is now commonplace to see the use of d.c. to d.c. converters providing both a step-up function for the positive supply rail while at the same time providing the negative supply rail for the output amplifier. The use of switched-mode supplies for the public address (PA) and high power amplifiers for guitars and rock concerts is becoming of greater significance. The subject of SMPS has previously been covered, but is mentioned because of the implications of the switching amplifier.

12.2 SWITCHING AUDIO POWER AMPLIFIERS

Switching audio power amplifiers have sometimes been referred to as digital amplifiers out of a sense of maintaining some connection with the computer generation or to appear more up to date perhaps and, although not completely providing an accurat description, none the less indicates that these amplifiers operate by having the output power transistors either completely on or off, i.e. they switch. (The switching amplifier requirement arises out of the very real improvement in efficiency which may be achieved.) This technique frequently uses pulse-width modulation to accomplish the necessary sound amplification with the loudspeaker sometimes being the sole performer of output signal averaging.

The full switched-mode power amplifier is the obvious choice within the context of improved efficiency. There is the other alternative which involves linear amplification coupled to switching in an additional supply rail(s) as increasing output power levels may warrant. This type of amplifier has been given the title of *cascade amplifier*, since the supply bus rails are switched in-circuit in a cascade configuration. Such an amplifier is shown in a very simplified format in Figure 12.4. The cascade amplifier operates as follows.

The input signal feeds a unity gain non-inverting power amplifier (power voltage follower) and this signal is compared against the two lower voltage supply rails. If the signal reaches to within a few percent of he value of either positive or negative supply rail, the appropriate comparator switches in the next higher supply rail of the corresponding polarity. This enables the lower value supply rails to be set such that the amplifier operates at 70% efficiency for 50% output power. The next higher value of supply rail is fixed to enable the amplifier to operate at 70% efficiency at full output power, thereby giving a worthwhile improvement in efficiency at medium levels of output power, an efficiency figure which is not achievable with a non-cascade amplifier.

The voltage ratings of Q1 and Q4 need only equate to the difference between V_{dd1} and V_{dd2} and between V_{ss1} and V_{ss2}. This requirement becomes self-evident when it is realised that Q1, when it is off, cannot be exposed to more than the

186 AUDIO AMPLIFICATION

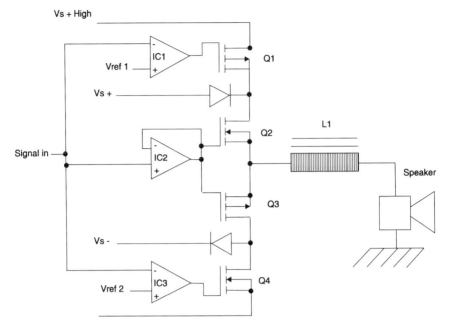

Figure 12.4 Simplified circuit A cascaded amplifier

difference in the value of the two positive supplies. Similarly, Q4 cannot be exposed to more than the difference between the two negative supplies. Both Q2 and Q3, however, should be rated for the full value of the sum of the voltages of V_{dd1} and V_{ss1}: where if $V_{dd1} = V_{ss1}$, then BV_{DSS} of Q2 and Q3 must be greater than $2V_{dd1}$.

A simplified block schematic of such a amplifier is shown in Figure 12.5. It should be noted that the audio signal must first be fed to a precision full-wave rectifier so that a unipolar signal is compared against the reference voltage in the pulse-width modulator. The PWM output is gated by the absolute value output of the precision full wave rectifier to ensure that the correct polarity supply is switched to the loudspeaker which in this case acts as part of the output filter. It should be borne in mind that the loudspeaker represents a non-linear load (of non-unity power factor) to the amplifier.

If the amplifier is meant to provide a signal reflecting the fidelity of the original sound input, the switch frequency will then have to be sufficiently high and this will require the disabling of the parasitic body drain diode of the output MOSFET. BJTs cannot operate satisfactorily at the frequencies being considered here.

It has already been pointed out that the primary interest in switching amplifiers stems from a need to improve efficiency. Non-cascaded linear amplifiers can operate at efficiencies up to 70% if they are correctly designed. This maximum figure occurs at maximum output power. Reducing output power, while

12.3 THE AUDIO POWER AMPLIVERTER

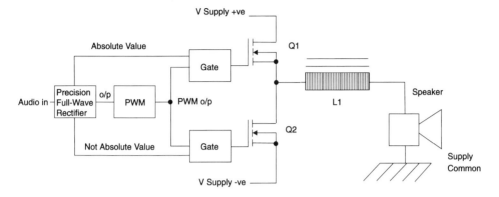

Figure 12.5 Block schematic of switching power amplifier

maintaining the output capability, results in reducing efficiency; for example, 50% efficiency is achievable at half output power level. The cascade amplifier gives a small though worthwhile boost to efficiency at the expense of increased cost and complexity.

The switching amplifier of Figure 12.5 provides an overall efficiency of up to 90% without taking into account power supply efficiency. If the power supply is significantly less efficient than 100% (e.g. much less than 97.5%), then the overall efficiency of the power amplifier reduces to being the product of the individual efficiencies. This is best typified in the car booster. The d.c. to d.c. converters probably operate at about 80–85% efficiency and if the output amplifier operates at maximum efficiency (70%), then the maximum overall efficiency equates to 56–59.5%.

If the output amplifier in the car booster was replaced by a switching stage with an efficiency of 80–85% then the overall efficiency improves to between 64% and 72.25%, the latter figure being quite creditable.

At output powers of 500 W even 72.25% is still found to be less than optimal when it is compared with an overall figure of up to 85%. This equates to a difference in dissipation of 192 W for an efficiency of 72.25% and 88 W for an efficiency of 85%.

Achieving the figure of 85% requires the amplifier and power supply to be configured into a single converter circuit. Such a circuit has the generic name of *ampliverter*.

12.3 THE AUDIO POWER AMPLIVERTER

Ampliverters may be configured utilising virtually any one of a number of power converter topologies in such a manner that the fundamental switch frequency

188 AUDIO AMPLIFICATION

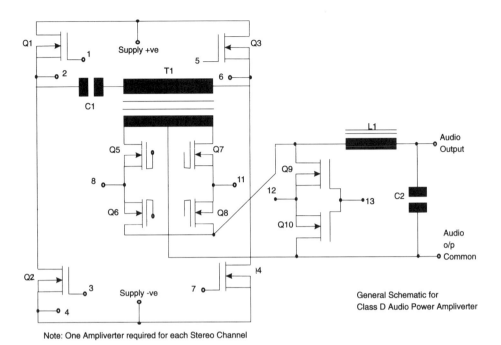

Figure 12.6 Simplified circuit of the ampliverter

is pulse-width modulated by the audio signal. This approach enables signal amplification and power conversion to take place at one and the same time by a single power stage. The overall efficiency quantification reduces to the efficiency of power conversion.

A simplified schematic which outlines the basic principles and concepts of an ampliverter is shown in Figure 12.6. The fundamental nature of the circuit of Figure 12.6 is an H-bridge d.c. to a.c. inverter followed by full-wave rectification which is performed by bi-directional synchronous rectifiers. The bi-directionality of the output synchronous rectifiers is a direct consequence of the non-linear characteristic of the loudspeaker as a load. The reactive nature of this load requires full bi-directionality in the output synchronous rectifiers. The output filter of L1 and C2 is required if the ampliverter is required to perform in either mid-fi or hi-fi applications.

The fundamental switch frequency (f_o) is determined by the level of Total Harmonic Distortion (THD) which has been specified for the output. It can be demonstrated that THD is inversely proportional to both the carrier frequency f_o and the modulating frequency f_m which in this case is the audio signal.

In low distortion applications the figure for f_o could well be several hundred kilohertz and consequently all of the precautions of EMI/RFI suppression will have to be observed, if the ampliverter is not to function as a high-powered

12.3 THE AUDIO POWER AMPLIVERTER

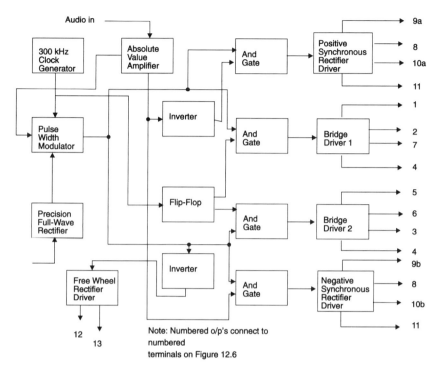

Figure 12.7 Schematic of control, gating and driver circuits for the power ampliverter of Figure 12.5

amplitude modulated (AM) transmitter in the long-wave region. It is also quite likely that the ampliverter may require containment within a fully screened enclosure to ensure that AM interference to nearby equipment does not occur.

The control circuit for the ampliverter can easily be configured out of standard building blocks which may utilise readily available proprietary integrated circuits. The full schematic of the control small-signal circuitry is given in Figure 12.7. The circuits, such as the absolute value amplifier and full wave rectifier, may be easily constructed from simple *operational amplifiers*. The two functions of rectification and absolute value extraction can be combined into one overall circuit.

The information to design all of the circuits which are enclosed within the relative block outlines is beyond the scope of this book. The information is none the less readily available in applications handbooks from the major suppliers of linear and digital integrated circuits. The reader is encouraged to become conversant with all of the most relevant small-signal circuit functions.

Appendix
Safe Operating Area Failures in BJTs

Within the text of the preceding chapters I have frequently referred to the potential break-down mechanism known as second break-down. This phenomenon, unfortunately, is an ever-present threat in BJTs. The failure mechanism is frequently misunderstood by users and I will endeavour to explain this mechanism in some detail.

Before I progress too far I would like to make the reader aware that within the normal course of operation the BJT is quite likely to sustain a form of break-down which is frequently mistaken for second break-down but is an entirely different mechanism. This phenomenon is sometimes called *field intensity break-down*. This break-down mechanism is associated with dynamic voltage stress encountered during turn-off whereas second break-down is a thermal phenomenon due to *current crowding*. Unfortunately, second break-down can manifest itself during the normal 'on state' condition and/or during the turn-off transition of the transistor. We will now examine both conditions.

I would also like to point out that the descriptions given in this appendix refer in totality to *NPN transistors*.

A.1 FORWARD BIASED SECOND BREAK-DOWN

During the conduction time of the transistor, and with the flow of high values of collector current above the maximum value specified for the device, lateral thermal instabilities across the face of the junction can give rise to localised areas of extremely high current densities.

A simplified diagram of the condition of current crowding/current hogging is depicted in Figure A.1.

These high current densities may give rise to severe localised heating which could alter (increase) the local forward current gain (h_{fe}) in such a way as to further increase the current density. The effect can be syndromic and possible melting and ultimate failure of the junction could be the result.

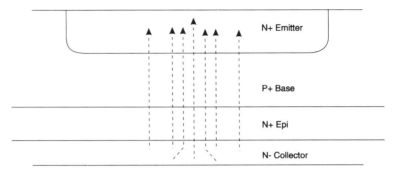

Figure A.1 Current flow at the onset of second breakdown

The only method which may be employed to ensure that this phenomenon does not manifest itself is to *never exceed the specified collector current rating as stipulated by the vendor.*

There are suppliers of BJTs who publish Over-load Safe Operating Areas for some of the transistors within their catalogue. Close examination of the do's and the don'ts for use of the O/L SOA and the small print relating to the permissible number of excursions into this area implies the virtual non-usefulness of this feature.

A.2 REVERSE BIAS SECOND BREAK-DOWN

If an inexperienced person were to carefully test any one of three parameters on a curve tracer relating to applied collector–emitter voltage they would observe that the break-down curve resembled the curve shown in Figure A.2. (Note that the use of a curve tracer for this test is not recommended by most suppliers of BJTs and is the reason for caution in the implementation of this test.)

The three parameters which may be examined are defined as follows:

(1) V_{CBO}. This parameter refers to the break-down voltage across the collector–base junction with the emitter terminal open-circuited. Purists would state that this parameter is the break-down voltage of a reverse-biased diode and therefore would imply that the break-down mechanism is avalanche of the junction. This proves to be the case and the break-down mechanism is therefore referred to a *primary break-down*. All other break-down mechanisms are referred to as *secondary break-down* or second break-down.

(2) C_{CER}, V_{CES}, V_{CEX}. I have placed these three parameters together since they all refer to the break-down voltage as measured from collector to emitter with the base terminal having different conditions as may be specified by the vendor. V_{CES} applies when there is a specific value of resistance connected externally from base to emitter, V_{CES} applies when the base is externally

A.2 REVERSE BIAS SECOND BREAK-DOWN

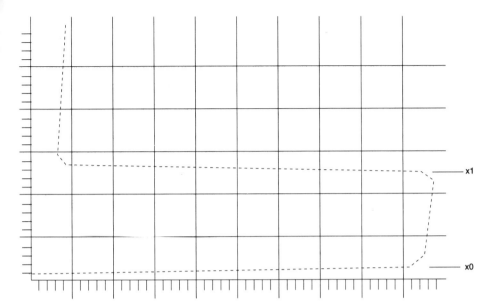

Figure A.2 Curve tracer wave form of V_{ce} break-down

shorted to the emitter and finally V_{CEX} applies when the base is reverse-biased with respect to the emitter by a specified voltage.

(3) V_{CEO} and $V_{CEO(SUS)}$. This parameter refers to the collector to emitter break-down voltage with the base terminal being open-circuited and measured for a specified value of current flow through the transistor. The suffix (sus) refers to the voltage which may be sustained for any period of time and usually implies a higher value of current than for the normal value of V_{CEO}.

I will not go into the exact details as to how the test should be conducted. The test which should be carried out is one of the VCE tests with the base not left open-circuited. The trace should be carefully observed on the display and the applied voltage increased slowly.

Provided the user proceeds with caution in carrying out the tests a curve similar to that given in Figure A.2 should be displayed on the screen of the curve tracer.

An explanation of the displayed curve is as follows:

(1) The vertical portion of the trace above point x_0 and below point x_1 is primary break-down and the transistor is not under any condition of distress. Provided the temperature is held to a safe level the device will sustain this condition indefinitely. Voltages within the die related to base and emitter will be found to indicate that emitter injection is not taking place and transistor action does not occur.

(2) Increasing the voltage only slightly above point x_1 starts the action referred to as emitter injection. This action is the result of the flow of two currents from the base region, i.e. an *electron current* from base to emitter and a *hole current* from base to collector. The hole current to the collector results in transistor action whereby an electron current now flows from collector to emitter via the base and constitutes the major part of the electron current initially alluded to as being one of the base currents. *Emitter injection* is said to have commenced. With the onset of emitter injection the reader will observe that the transistor can no longer support any significant voltage across its structure and V_{CE} is seen to collapse. If the impedance in the collector circuit was sufficiently small the collector current I_C would be completely out of control and would climb to a dangerous level with consequential destruction to the device in question. The initiation of the collapse of V_{CE} and the dramatic increase in I_C indicates that the device has entered the region of second break-down.

The tests conducted so far relate to the transistor initially being exposed to primary break-down with the potential to enter second break-down if care is not exercised. Note the voltage at point x_0 and refer to it as point $x_{0(bc)}$. The reason for noting this value will become clear in due course.

The test should be repeated for the transistor now having its base open-circuited and with the measurement being made for the voltage and current conditions applicable to (1) above between points x_0 and x_1 only. Note this second value of x_0 which should be referred to as point $x_{0(bo)}$. There is a difference between the values of $x_{0(bc)}$ and $x_{0(bo)}$.

A.3 REVERSE BIAS SAFE OPERATING AREA (RBSOA)

If the reader refers to the data sheet of the transistor which has been used for the measurements a curve similar to the one given in Figure A.3 will be noted and should be found to demonstrate the Reverse-Bias Safe Operating Area (RBSOA). It will be noted that the curve is not rectangular but will probably have two nearly perpendicular portions at two different voltages. These perpendiculars are marked on Figure A.3 as $V_{CEO(sus)}$ and V_{CBO} respectively.

A complete understanding of the ramifications of the RBSOA curve will demonstrate the need for *switching aid networks* (or *current snubbers* as they are frequently referred to). These networks are a requirement to ensure that the 'turn-off' *switching trajectory* always remains within the confines of the RBSOA. Note that current snubbers are frequently used in order to reduce the switch-off losses in BJTs, but in this instance the snubber is performing a secondary task in addition to maintaining the turn-off behaviour of the device within the bounds of the RBSOA.

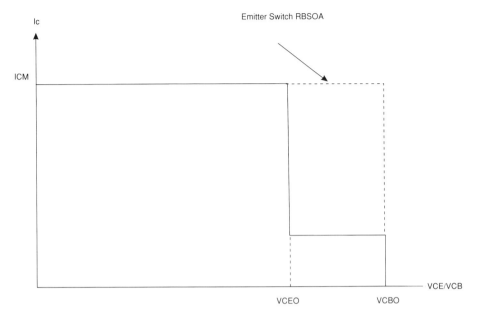

Figure A.3 Reverse-bias safe operating area curve

An interesting aside is the fact that emitter switching of the BJT rounds out the RBSOA curve to almost occupy a rectangle bounded by V_{CBO} and I_{CM}. The inference is that emitter switching removes the threat of reverse bias second break-down during turn-off.

The reader is reminded that the MGT's safe operating area is completely rectangular and that current snubbers are not essential. They may, on the other hand, be entirely desirable in reducing the turn-off losses within the device.

The reason to stress that second break-down is a thermal effect is that in most high voltage circuits the collector current is usually curtailed, except in the case of a saturated transformer, and that failure of the BJT, repeatedly exposed to second break-down, is the result of the thermal stresses.

A.4 FIELD INTENSITY BREAK-DOWN

The likelihood of another break-down mechanism referred to as Field Intensity Break-down (FIB) was indicated earlier. An explanation of this phenomenon requires the reader to examine the two diagrams in Figure A.4.

In Figure A.4(a) the broken line extending from the P-base region into the N region of the collector represents normal *base depletion* when the transistor is on and in a saturated condition. The depth to which depletion occurs depends

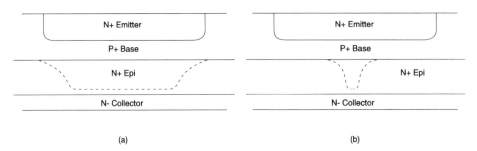

Figure A.4 Conditions relating to field intensity break-down within BJT

entirely upon the design of silicon structure itself. The thickness of silicon from the bottom of the *depletion trough* to the collector connection is only a few micrometres but this is of little consequence since the collector–emitter voltage and by implication the collector–base voltage is small.

Consider the effects as turn-off is initiated especially if the transistor has been installed in a circuit with an inductive load. This is depicted in Figure A.4(b). The depth of the depletion trough remains almost unchanged. All that has happened is that charge removal during *storage time* has narrowed the dimension between the walls of the trough. The collector voltage could very well have reached a maximum and the situation is that the thickness of the silicon from the bottom of the trough to the collector connection is insufficient to support the applied voltage. In this instant the likelihood of an arc occurring is extremely high. The arc is initiated by the intensity of the electric field. This break-down mechanism is a once-only action. The transistor gets no second chance.

The design of the transistor should ensure that this situation cannot occur. If the possibility exists that failure in the transistor is due to FIB then the only recourse is to heavily *snub* the transistor if the vendor is not prepared to modify the design structure of the device.

Index

Active clamp *see* Clamp
Amplifier
  booster 178
  cascaded 179
  switch-mode 178
Ampliverter 180
Avalanche 45
  see also Break-down

Ballast
  electronic 162
Bias
  forward 183
Bootstrap 37
Breakdown
  field intensity 18, 314
  secondary (second) 183

Charge control 3
Clamp
  Active 43
  capacitive 51, 54
  passive 46
  topological 60
Conductivity modulation 10
Connection
  cascode 98, 101
    parallel 82, 107, 173
  series 101
Control
  charge 3
  gate 23, 176
  slave 101
Controllers (Motor)
  brushed DC 135, 148
  brushless DC 139
  reluctance 140

  shaded pole 143
  stepper 151
  universal 141
Converter
  buck 125
  bridge 120
  cscaded 125
  fly-back 117, 121
  fly-forward 123
  push–pull 119, 125
  quasi-resonant 130
Current
  limit 65
  measurement 67
  mirror 62
  sense 65, 176
  shunt 69
  snubbers
    turn-on 71
    turn-off 46
  trip 67

Damage, electrostatic 18
Decoupling 87
Detection, linear 68

Effects
  MGT structures 5
  package outline 78
Emitter switching 98
Emitter injection 148

Fall time
  current 14
  voltage 14
Field cancellation 87
Fuel injection 148

Ignition 151
Interference
 conducted 86, 164
 EMI/RFI 84
 radiated 86, 164
Interference reduction 86

Latching 16
Layout 59, 87

Migration, ion 19

Parallel connection 82, 173
Power supply
 battery charger 142
 switch-mode 116
  see also Regulators, Switching
 uninterruptible (UPS) 97
Protection
 Over current 61
 Over voltage 61
Punch through 19

Ratio, sense 64
Regulators
 linear
  Series 111
  Shunt 111
 switching 116

Resistor
 voltage dependent 56

Safe Operating Area
 Reverse Bias (RBSOA) 119, 187
Snubber
 current 46
 strays 9, 91
 turn-on 71
Suppresor
 voltage transient 55
Switch
 high side 159
 low side 159
Switching aid, IGT 107
Switching times 15
 current 41
 voltage 26, 42

Thickness, gate oxide 18
Time, delay 15
Transistor
 insulated gate (IGT) 11
 MOSFET, lateral 22
 MOSFET (mirror), current sensing 35, 63, 177
Turns Ratio Variation 126